World Scientific Advanced Series in Dynamical Systems

Vol. 3

INVITATION TO C*– ALGEBRAS AND TOPOLOGICAL DYNAMICS

Jun Tomiyama

World Scientific
Singapore • New Jersey • Hong Kong

ADVANCED SERIES IN DYNAMICAL SYSTEMS

Editor-in-Chief: K. Shiraiwa

Vol. 1: **Dynamical Systems and Nonlinear Oscillations**
edited by G. Ikegami

Vol. 2: **Dynamical Systems and Singular Phenomena**
edited by G. Ikegami

Vol. 3: **Invitation to C*-algebras and Topological Dynamics**
edited by J. Tomiyama

Forthcoming:

Vol. 4: **Numerical Methods for Chaotic Dynamical Systems**
edited by H. Kawakami

World Scientific Advanced Series in Dynamical Systems

Vol. 3

INVITATION TO C*– ALGEBRAS AND TOPOLOGICAL DYNAMICS

Jun Tomiyama

World Scientific

Singapore • New Jersey • Hong Kong

Published by

World Scientific Publishing Co. Pte. Ltd.
P.O. Box 128, Farrer Road, Singapore 9128

U.S.A. office: World Scientific Publishing Co., Inc.
687 Hartwell Street, Teaneck NJ 07666, USA

Library of Congress Cataloging-in-Publication data is available.

INVITATION TO C*-ALGEBRAS AND TOPOLOGICAL DYNAMICS

ISBN 9971-50-338-7

Printed in Singapore by Fong and Sons Printers Pte. Ltd.

Preface

Measure theoretical dynamics has a long history of the interplay with theory of von Neumann algebras. The result are fruitful and there are substantial mutual contributions. The counterpart of this aspect for the theory of operator algebras and dynamics is the interplay between topological dynamics and theory of C*-algebras. It is coming to be very promising and interesting, though at present particularly topological dynamical systems are providing attractive materials to the theory of C*-algebras. It is however only recent years that this developing area has been drawing wide attentions.

This book is designed to introduce some aspects of this interesting field in a way as elementary as possible but including some quite recent results. The readers need no preparation for C*-algebras but a little more knowledge about topological dynamics rather than the contents of the first chapter would help them to understand main ingredients of the text well.

The author is deeply indebted to S. Kawamura for stimulating discussions and careful reading of the manuscript during the presentation of this book.

Tôkyô,
February 1987

Jun Tomiyama

CONTENTS

INTRODUCTION

Let X be a compact Hausdorff space. In principle, to consider the structure of the space X is equivalent to studying the algebra of all complex valued continuous functions on X, $C(X)$. Suppose further that X admits a homeomorphism σ or a diffeomorphism σ if X admits further differential structure such as manifolds. The map σ reflects the algebra $C(X)$ as an automorphism α with $\alpha(\bar{f}) = \overline{\alpha(f)}$ (*-automorphism). This means that we are concerned with an action α of the integer group Z. Thus, to consider the topological dynamical system $\Sigma = (X, \sigma)$ comes to be equivalent to considering the system $\{C(X), \alpha, Z\}$. With usual supremum norm the algebra $C(X)$ is regarded as a Banach *-algebra of special kind called a (commutative) C^*-algebra. Let H be a Hilbert space and $B(H)$ be the algebra of all bounded linear operators on H. Now suppose that there is a faithful representation π of $C(X)$ on H as a self-adjoint subalgebra of $B(H)$ such that there exists a unitary operator u with $\pi(\alpha(f)) = u\pi(f)u^*$. The system $\{C(X), \alpha\}$ is then transfered to the covariant system $\{\pi(C(X)), u\}$ and finally it will be transplanted on the C^*-algebra $C^*(\pi(C(X)), u)$ generated by $\pi(C(X))$ and u. This C^*-algebra becomes necessarily non-commutative unless the automorphism α is trivial. One should then choose a good covariant representation $\{\pi, u\}$ so that it generates the algebra reasonably well. By the covariant condition that $\pi(\alpha(f)) = u\pi(f)u^*$, $C^*(\pi(C(X)), u)$ is regarded as the norm closure of the self-adjoint linear subspace consisting of those elements $\sum_{k=-n}^{n} \pi(f_k)u^k$.

Thus, we ask for condition of generation that the set $\{u^n | n \in Z\}$ should be independent over the algebra $\pi(C(X))$ and moreover at least for norm $\| \sum_{k=-n}^{n} \pi(f_k)u^k \| \geq \| \pi(f_0) \|$. In this way, we finally reach the notion of the C*-crossed product $A_\Sigma = C(X) \rtimes_\alpha Z$ (and its reduced form $C(X) \rtimes_{\alpha r} Z$) called the transformation group C*-algebra. Therefore, again in principle, from the C*-algebraic point of view the study of a topological dynamical system $\Sigma = (X, \sigma)$ is recognized as the study of the associated transformation group C*-algebra A_Σ. We may generalize the situation from single homeomorphism to a (topological) group of homeomorphisms acting on a locally compact space X. Of course, there must be basic distinction in the way of understanding of these objects. We either consider them as the objects for topological dynamics or as those for C*-algebras. Thus naturally, actual interplays between these two kinds of objects are not proceeding in an equal way. The theory of C*-algebras has been much benefitted from topological dynamical systems as its driving materials as in the case of the C*-algebra A_θ arised from an irrational rotation of the torus. However there have been considerable developments in the field recently as we shall see some of them in this book.

In the setting of measure theoretical dynamical systems such as $\{L^\infty(X, \mu), \rho\}$ for a non-singular ergodic transformation ρ, the associated algebras of operators in B(H) become crossed products (such as $L^\infty(X, \mu) \rtimes_\alpha Z$) of von Neumann algebras. In this category, dynamical systems provide a great deal of examples of factors as well as experimental field for the theory, but there are also big contributions from the side of von Neumann algebras to the ergodic theory for not-measure-preserving ergodic transformations.

One may also hope for further fruitful interplays between topological dynamics and theory of C*-algebras (together with that of von Neumann algebras) and expect mutual contributions as well.

It is to be noticed that references attached are rather optional, though they are directly concerned with the contents of the book. The author does not intend to present a general list of references related to topological dynamics and transformation group C^*-algebras.

CHAPTER 1

TOPOLOGICAL DYNAMICAL SYSTEMS

We present here basic concepts and results in the theory of topological dynamics for later use. In general, in the C*-algebraic context the setting for dynamical systems is provided by a locally compact space X with a locally compact transformation ~~group G~~ on X. Since however most relevant things will appear, in principle, in a discrete topological dynamical system (X, σ) we simplify our presentation to the case of a topological dynamical system (X, σ) for a compact space X, though we shall partly discuss in Chaps. 3 and 4 about the action of a discrete group.

1.1 Minimality and Ergodicity of Topological Dynamical Systems

Throughout this chapter we denote a topological dynamical system by $\Sigma = (X, \sigma)$ or simply by Σ where the space X is always a compact Hausdorff space and σ is a homeomorphism. The algebra $C(X)$ means the algebra of all complex valued continuous functions on X with the supremum norm $\|f\| = \max|f(x)|$. The set $0_\sigma(x)$ for a point x in X means the orbit of x by σ, that is,

$$0_\sigma(x) = \{\sigma^n x \mid n \in Z\} \quad .$$

We denote by $0_\sigma^+(x)$ the positive orbit of x,

$$0_\sigma^+(x) = \{\sigma^n x \mid n = 0,1,2,\dots\} \quad .$$

The closure of these sets are written as $\overline{O_\sigma(x)}$ and $\overline{O^+_\sigma(x)}$. We abbreviate σ in the above notations when no confusion arises.

Definition 1.1.1. A topological dynamical system $\Sigma = (X, \sigma)$ is said to be minimal if there exists no proper closed set A in X such that $\sigma(A) \subset A$. The system Σ is said to be recurrent if $x \in \overline{O^+_\sigma(\sigma x)}$ for every point x.

There are variants of the definition of minimality as in the following:

Proposition 1.1.1. The following conditions for a dynamical system Σ are equivalent;

(1) Σ is minimal;

(2) $\overline{O^+_\sigma(x)} = X$ for every point x;

(3) $\overline{O_\sigma(x)} = X$ for every point x;

(4) There is no proper closed set A in X with $\sigma(A) = A$.

Proof. Since the set $\overline{O^+_\sigma(x)}$ is not empty and satisfies that $\overline{\sigma(O^+_\sigma(x))} \subset \overline{O^+_\sigma(x)}$ we have the assertion (2) if Σ is minimal. The implication (2) $\Rightarrow$ (3) $\Rightarrow$ (4) are trivial. Assume the condition (4) and let A be a non-empty closed set such that $\sigma(A) \subset A$. We then have a decreasing sequence of closed sets $\{\sigma^n(A) \mid n = 0,1,2,\ldots\}$. Since X is compact the intersection $B = \bigcap_{n=0}^{\infty} \sigma^n(A)$ is a nonempty closed set and $\sigma(B) = B$. Hence $B = X$ and the set A coincides with X. Thus the system Σ is minimal. This completes all proofs.

Example 1.1.1. Let θ be an irrational number of the torus T and define the homeomorphism σ_θ: $t \to t + \theta$ (mod 1) in T. Then the system (T, σ_θ) is minimal. Since the map σ_θ naturally induces a roation $\hat{\sigma}_\theta$; $e^{2\pi it} \to e^{2\pi i(t+\theta)}$ with irrational angle θ in the unit circle S^1 this homeomorphism σ_θ (together with $\hat{\sigma}_\theta$) is often called an (rigid) irrational rotation. We shall show the minimality of the system.

We show first that for any integer $l \in N$ there exists positive integers m and n such that $|m\theta - n| < \frac{1}{\ell}$. In fact, divide the interval $[0, 1)$ into ℓ-pieces;

$$[0, 1) = [0, \tfrac{1}{\ell}) \quad [\tfrac{1}{\ell}, \tfrac{2}{\ell}) \ \dots \ [\tfrac{\ell-1}{\ell}, 1) \qquad .$$

If we consider $\ell + 1$ points $\{0, \theta - [\theta], 2\theta - [2\theta], \dots, \ell\theta - [\ell\theta]\}$ where $[x]$ means the integer part of x, some two points $p\theta - [p\theta]$ and $q\theta - [q\theta]$ must be in an interval $[\frac{j}{\ell}, \frac{j+1}{\ell})$. Hence

$$|(p - q)\theta - ([p\theta] - [q\theta])| = |p\theta - [p\theta] - (q\theta - [q\theta])| < \frac{1}{\ell} \qquad .$$

It suffices to put $m = p - q$ and $n = [p\theta] - [q\theta]$ assuming $p > q$. We assert next that the orbit of θ is dense in T. Indeed, take $y \in T$ and a positive ε with $0 < \varepsilon < 1 - y$. Choose positive integers 1, m and n such that $|m\theta - n| < \frac{1}{\ell} < \varepsilon$. If $\alpha = m\theta - n > 0$ we consider a positive integer k such that $k\alpha \leq y \leq (k + 1)\alpha$. Then

$$|y - k\alpha| < \alpha < \frac{1}{\ell} < \varepsilon$$

and $k\alpha \equiv km\theta \bmod 1$. When $\alpha < 0$, one may similarly find an integer k for $-\alpha$ such that

$$|y - k(-\alpha)| < -\alpha < \varepsilon \qquad .$$

Thus, $O_{\sigma_\theta}(\theta)$ is dense in T and this easily implies the minimality of (T, σ_θ).

We shall be deeply concerned with this example later from the point of view of the theory of C*-algebras.

Now the above proposition shows that if the system is minimal it is recurrent. For the converse we have

<u>Example 1.1.2.</u> $\rho_\theta(t) = t + \theta$ in T for a rational number θ. In this case the map ρ_θ is apparently a periodic homeomorphism with

period p for $\theta = \frac{q}{p}$ (irreducible fraction). Hence ρ_θ is not minimal but it is recurrent.

Example 1.1.3. Let σ be the map $\sigma(x) = x^2$ on the unit interval $[0, 1]$. The system $([0, 1], \sigma)$ is neither minimal nor recurrent. The only invariant minimal closed sets are $\{0\}$ and $\{1\}$.

The following theorem shows that any topological dynamical system contains a minimal dynamical system.

Theorem 1.1.2. Let $\Sigma = (X, \sigma)$ be a topological dynamical system. There exists then a nonempty closed set X_0 such that $\sigma(X_0) = X_0$ and the restricted dynamical system $\Sigma_0 = (X_0, \sigma|X_0)$ is minimal.

Proof. Let $\mathscr{F}$ be the collection of all nonempty closed sets A's such that $\sigma(A) \subset A$. Obviously, $\mathscr{F}$ is a nonempty family and we consider an order in $\mathscr{F}$ by inclusion. Let $\mathscr{F}_0$ be a decreasing chain of $\mathscr{F}$, then by the compactness of X the intersection $Y_0 = \bigcap_{A \in \mathscr{F}_0} A$ is not empty. The set Y_0 is clearly a lower bound of $\mathscr{F}_0$. Hence by Zorn's lemma we can find a minimal closed set X_0 in $\mathscr{F}$. Now the inclusion $\sigma(X_0) \subset X_0$ implies that $\sigma^2(X_0) \subset \sigma(X_0)$, which means that $\sigma(X_0) \in \mathscr{F}$. Hence by the minimality of X_0, $\sigma(X_0) = X_0$, and by definition of X_0 the system $\Sigma_0 = (X_0, \sigma|X_0)$ is minimal.

Definition 1.1.2. A dynamical system $\Sigma = (X, \sigma)$ is said to be topologically transitive if for nonempty open sets U and V there exists an integer n such that $\sigma^n U \cap V \neq \phi$.

Example 1.1.4. The well known symbolic dynamical system $(X(k), \sigma_k)$ of Bernoulli shift is topologically transitive. Here the space $X(k)$ is the product space of infinite copies of the finite set of integers $\{0,1,2,\ldots,k-1\}$ and the shift transformation σ_k is defined as

$$\sigma_k(x) = y \quad \text{for} \quad x = (x_n) \quad \text{and} \quad y = (y_n) \quad \text{with} \quad x_{n-1} = y_n .$$

It is clear that σ_k is one-to-one. The continuity of σ_k and σ_k^{-1} is seen from the following observation. Namely, the product topology of $X(k)$ is defined by the cylinder sets. Consider an ℓ-tuple of fixed coordinates $(n_1, n_2,\ldots,n_\ell)$ and let A be the set consisting of ℓ-tuples of subsets of $\{0,1,2,\ldots,k-1\}$. Then a cylinder set is defined as

$$C(n_1, n_2,\ldots,n_\ell) = \{x \quad X(k) \mid (x_{n_1}, x_{n_2},\ldots,x_{n_\ell}) \in A\}$$

$$= \bigcup_{(n_1, n_2,\ldots,n_\ell)\in A} \bigcup_{j=1}^{\ell} \{x \in X(k) \mid x_{n_j} = n_j\} \quad .$$

For this set C,

$$\sigma_k^{-1}(C) = \bigcup_{(n_1, n_2,\ldots,n_\ell)\in A} \bigcup_{j=1}^{\ell} \{x \in X(k) \mid x_{n_j+1} = n_j\}$$

is also a cylinder set as well as the set $\sigma_k(C)$. Hence σ_k is a homeomorphism.

We note that the topology of $X(k)$ is also defined by the metric:

$$d(x, y) = \begin{cases} 0 & \text{if} \quad x = y \\ \dfrac{1}{1 + \min\{|n|, x_n \neq y_n\}} & \text{if} \quad x \neq y \end{cases} .$$

Therefore if we regard the points of $X(k)$ as k-adic expansions of the number in the interval $[0, 1]$ the point x_0 whose expansion contains any expansion of such a number with finite length in its certain interval of coordinates satisfies the condition $\overline{O_{\sigma_k}(x_0)} = X(k)$. Thus the system is topologically transitive in a more strong sense.

Theorem 1.1.3. Consider the following four conditions for a dynamical system $\Sigma = (X, \sigma)$:

(1) There exists a point in X with dense orbit;

(2) If a proper closed set A satisfies the condition $\sigma(A) = A$, then the complement A^c is dense in X;

(3) Σ is topologically transitive;

(4) The set $\{x \in X, \overline{O(x)} \neq X\}$ is of first category.

Then, (1) $\Rightarrow$ (2) $\Rightarrow$ (3) and (4) $\Rightarrow$ (1). If X is second countable all conditions are equivalent.

Proof. (1) $\Rightarrow$ (2). Suppose that the set A with the condition (2) contains a nonempty open set U. Let x_0 be a point with $\overline{o(x_0)} = X$. There exists then an integer n such that $\sigma^n x_0 \in U \subseteq A$, and $o(x_0) \subseteq A$. Hence, $\overline{o(x_0)} \subseteq A \subseteq X$, a contradiction.

(2) $\Rightarrow$ (3). The set $B = \bigcup_{n=-\infty}^{\infty} \sigma^n(U)$ is obviously an open set with $\sigma(B) = B$. Hence by condition (2) B is dense in X and $B \cap V \neq \phi$. The converse is clear.

Now suppose that X is second countable and let $\{U_n | n = 1,2,3, \ldots\}$ be a countable base for a topology of X. From the assumption (3) the set $\bigcup_{k=-\infty}^{\infty} \sigma^k(U_n)$ is dense in X for every n, and its complement $\bigcap_{k=-\infty}^{\infty} \sigma^k(U_n^c)$ is a closed set with no interior point. On the other hand we have the following equivalences:

$$\overline{o(x)} \neq X \Leftrightarrow \text{there exists a set } U_n \text{ such that}$$

$$\overline{o(x)} \cap U_n = \phi \quad ,$$

$$\Leftrightarrow \text{there exists a set } U_n \text{ such that}$$

$$\sigma^k x \in U_n^c \text{ for every } k \quad ,$$

$$\Leftrightarrow x \in \bigcap_{k=-\infty}^{\infty} \sigma^k(U_n^c) \text{ for some } U_n$$

Thus

$$\{x \in X | \overline{o(x)} \neq X\} = \bigcup_{n=1}^{\infty} \bigcap_{k=-\infty}^{\infty} \sigma^k(U_n^c) \quad ,$$

which is of first category. This finishes the implication (3) $\Rightarrow$ (4). The assertion (4) $\Rightarrow$ (1): Since a compact space is not of first category by the category theorem there exists a point x in X with dense orbit.

In the last chapter we shall discuss about C*-algebras arising from topologically transitive dynamical systems. As for the spaces which admit such dynamical systems the unit interval does not admit such systems, for any homeomorphism on the unit interval becomes a monotone function so that every orbit cannot be dense in the interval. On the other hand, there are many topologically transitive systems on the torus such as irrational rotations. The 2-dimensional torus T^2 admits a topologically transitive dynamical system. For instance the homeomorphism defined by the matrix $\begin{pmatrix} 1 & 2 \\ 1 & 1 \end{pmatrix}$ is known to be topologically transitive but this fact is not trivial. On the sphere S^3 it is known (cf. Ref. 9) that there exists a diffeomorphism σ such that the system (S^3, σ) is minimal. Let (X, σ) be a topologically transitive dynamical system. Then an invariant continuous function on X is necessarily constant, but a system need not be topologically transitive even if there is no invariant functions except constant functions. In fact, consider a topologically transitive dynamical system (Y, σ) with a fixed point y_0 such as the case for T^2 mentioned above and let $(X, \hat{\sigma})$ be the two whole copies of (Y, σ) attached by identifying the point y_0. This dynamical system is apparently not topologically transitive, whereas an invariant function $f(x)$ on X is constant on each copy of Y in X and since copies are attached at the point y_0 the function $f(x)$ must be constant.

Let $(X, \mathscr{B}, \mu)$ be a probability measure space and recall that a map ρ on X is said to be measurable if $\rho^{-1}(A) \in \mathscr{B}$ for every set $A \in \mathscr{B}$. The system $(X, \mathscr{B}, \mu, \rho)$ may be called a measure theoretic dynamical system. The map ρ is said to be ergodic (with respect to μ) if $\rho^{-1}(A) = A$, $A \in \mathscr{B}$ implies either $\mu(A) = 0$ or $\mu(A^c) = 0$. In this case, μ is called an ergodic measure. If ρ is measure preserving, an ergodic measure is an extreme point in the set of all invariant probability measures. We call a map ρ an automorphism if it is a bimeasurable, one-to-one and onto map. An automorphism is further called a non-singular map if $\mu(A) = 0$ is equivalent to $\mu(\rho^{-1}(A)) = 0$. In this case, ρ induces an automorphism α of the algebra $L^{\infty}(X, \mu)$ defined by $\alpha(f)(x) = f(\rho^{-1}(x))$. This corresponds to the automorphism α defined in the algebra $C(X)$ by $\alpha(f)(x) = f(\sigma^{-1}x)$ for a topological dynamical system $\Sigma = (X, \sigma)$.

Proposition 1.1.4. Suppose that a topological dynamical system $\Sigma = (X, \sigma)$ is minimal, then there exists an invariant ergodic measure whose closed support is X.

Proof. Consider the automorphism α of $C(X)$ induced by σ. Let β be the transpose map of α in the dual space of $C(X)$, $M(X)$ regarded as the space of all Radon measures on X. Write

$$K = \{\mu \in M(X) \mid \mu \geq 0, \ \|\mu\| = \mu(X) = 1\} \quad .$$

The set K is a convex weak * compact set, which is invariant by β, whence by β^n. Therefore by the fixed point theorem of Kakutani-Markov we may find a fixed point of β in K. Furthermore the set of all fixed points is again a convex weak * compact set, hence it has an extreme point by the Krein-Milman theorem. This fixed point satisfies all conditions.

The existence of an invariant ergodic measure assures in some cases the existence of sufficiently many points whose orbits are dense in X. Namely we have

Proposition 1.1.5. Let X be a compact metric space and let μ be an invariant ergodic measure for a topological dynamical system (X, σ). If $\mu(U) > 0$ for every nonempty open set U, then

$$\mu(\{x \in X \mid \overline{o(x)} = X\}) = 1 \quad .$$

Proof. Let $\{U_n \mid n = 1,2,3,\ldots\}$ be a countable base of open sets in X. For each n the set $E_n = \bigcap_{k=-\infty}^{\infty} \sigma^k(U_n^c)$ is a Borel set with $\sigma(E_n) = E_n$. Moreover as in the proof of the implication $(3) \Rightarrow (4)$ in (1.1.4)

$$\bigcup_{n=1}^{\infty} E_n = \{x \in X \mid \overline{o(x)} \neq X\} \quad .$$

Since μ is ergodic, $\mu(E_n)=0$ or $\mu(E_n)=1$. But as $\mu(U_n)>0$ we see that $\mu(E_n^c)>0$ and $\mu(E_n)=0$. Hence we get the conclusion.

Definition 1.1.3. a) A dynamical system $\Sigma=(X,\sigma)$ is said to be uniquely ergodic if there exists a unique invariant ergodic probability measure in X. b) The system Σ is said to be strictly ergodic if it is minimal and uniquely ergodic.

For a function f on X we write

$$f_n(x) = \frac{1}{n}\sum_{k=0}^{n-1} f(\sigma^{-k}x) \quad .$$

c) A point x in X is said to be generic for a measure μ if

$$f_n(x) \to \mu(f) \qquad \text{for every } f \in C(X) \quad .$$

We shall treat many examples of uniquely ergodic dynamical systems and strictly ergodic dynamical systems later. On the other hand we recall that in the case of an ergodic measure theoretic dynamical system $(X, \mathscr{B}, \mu, \sigma)$ the averaging function $f_n(x)$ converges to a constant function almost everywhere for every μ-integrable function f.

Theorem 1.1.6. Let μ be an invariant probability measure in X for a dynamical system $\Sigma=(X,\sigma)$. Then the following conditions are equivalent;

(1) Σ is uniquely ergodic with a unique invariant measure μ;
(2) $f_n(x)\to\mu(f)$ uniformly for every continuous function f;
(3) Every point x of X is generic for μ.

Proof. The implication (2)$\Rightarrow$(3) is trivial. The assertion (1)$\Rightarrow$(2). Consider the subspace

$$S = \{f-\alpha(f) \mid f\in C(X)\} \quad .$$

From the assumption the null space of S in $M(X)$, S^0, is one dimensional, that is, $S^0=\{\mathbb{C}\mu\}$. The bipolar S^{00} in $C(X)$ which is the null space of μ coincides with the norm closure of S. Since $f - \mu(f)$ belongs to the set S^{00} for every $f \in C(X)$, for any positive number ε there exists a function g in $C(X)$ such that

$$\| f - \mu(f) - (g - \alpha(g)) \| < \varepsilon \quad .$$

Now, as

$$f_n(x) = \frac{1}{n} \sum_{k=0}^{n-1} f(\sigma^{-k}(x)) = \frac{1}{n} \sum_{k=0}^{n-1} \sigma^k(f)(x)$$

we have

$$\| f_n - \mu(f) - \frac{1}{n}(g - \alpha^n(g)) \|$$

$$= \frac{1}{n} \sum_{k=0}^{n-1} \alpha^k(f) - \frac{1}{n} \sum_{k=0}^{n-1} \mu(f) - \frac{1}{n} \sum_{k=0}^{n-1} \alpha^k(g - \alpha(g)) \|$$

$$\le \frac{1}{n} \sum_{k=0}^{n-1} \| \alpha^k(f - \mu(f) - (g - \alpha(g))) \|$$

$$= \| f - \mu(f) - (g - \alpha(g)) \| < \varepsilon \quad .$$

Namely,

$$\| f_n - \mu(f) - \frac{1}{n}(g - \alpha^n(g)) \| < \varepsilon$$

for every positive number n.

It follows that

$$\overline{\lim_{n \to \infty}} \| f_n - \mu(f) \| \le \varepsilon \quad .$$

Thus, the function $f_n(x) - \mu(f)$ converges to 0 uniformly. For the assertion (3) $\Rightarrow$ (1) let ν be an invariant probability measure. We shall show that $\mu = \nu$. By the assumption, $f_n(x)$ converges to $\mu(f)$ for every x. On the other hand,

$$\| f_n \| \leq \frac{1}{n} \sum_{k=0}^{n-1} \| \alpha^k(f) \| = \| f \| \quad .$$

Hence by the Lebesque convergence theorem

$$\mu(f) = \int_X \mu(f) d\nu = \lim_{n\to\infty} \int f_n(x) d\nu$$

$$= \lim_{n\to\infty} \frac{1}{n} \sum_{k=0}^{n-1} \int \alpha^k(f)(x) d\nu$$

$$= \lim_{n\to\infty} \frac{1}{n} \sum_{k=0}^{n-1} \nu(\alpha^k(f)) = \nu(f) \qquad f \in C(X) \quad .$$

Namely the measure ν coincides with μ. This completes all proofs.

We remark that the system in the example 1.1.3 is not uniquely ergodic. Indeed, by definition of σ we have $f_n(0) = f(0)$ and $f_n(1) = f(1)$ for every function f in $C(X)$ so that if $f(0) \neq f(1)$ the function $f_n(x)$ cannot converge to a constant function, contradicting the condition (3) in the above theorem.

Proposition 1.1.7. Suppose that the dynamical system $\Sigma = (X, \sigma)$ is uniquely ergodic with the unique measure μ. Then Σ is minimal if and only if $\mu(U) > 0$ for every nonempty open set U.

Proof. Suppose that Σ is minimal and $\mu(U) = 0$ for a nonempty open set U. The set $\bigcap_{k=-\infty}^{\infty} \sigma^k(U^c)$ is a nonempty closed invariant set. Hence, by the minimality of the system, the set coincides with X, which implies that $U = \phi$, a contradiction.

Conversely assume the condition for μ and that Σ were not minimal, there exists then a proper closed set A with $\sigma(A) = A$. We apply the same argument as in the proof of (1.1.5) and obtain a σ-invariant probability measure μ_A on A. Put $\mu'(B) = \mu_A(B \cap A)$ for every Borel set B of X. The measure μ' is clearly an invariant probability measure and $\mu'(A^c) = 0$. Hence $\mu' \neq \mu$, a contradiction.

1.2 Equicontinuous Dynamical Systems and Ellis Semi-groups

Recall that a compact space X has a unique uniform structure which describes the original topology. With this structure a continuous function on X becomes always uniformly continuous. We denote $\mathcal{U} = \{\alpha\}$ the family of indexes on X which defines the uniform structure of X. Namely, $\mathcal{U}$ is a filter of open sets in the product space $X \times X$ satisfying the following conditions;

(a) Every index α contains the diagonal set Δ;

(b) If $\alpha = \{(x, y)\} \in \mathcal{U}$, then $\alpha^{-1} = \{(y, x) | (x, y) \in \alpha\} \in \mathcal{U}$;

(c) For each index α there exists an index β such that $\beta^2 \subset \alpha$, where β^2 is defined as the set of $X \times X$, $\{(x, z)\}$ for which there exists $y \in X$ with $(x, y) \in \beta$ and $(y, z) \in \beta$.

For each point $x \in X$ the sets

$$\{\alpha(x) | \alpha \in \mathcal{U}\} = \{\{y \in X, (x, y) \in \alpha\} | \alpha \in \mathcal{U}\}$$

form a base of neighborhoods of x in X.

Definition 1.2.1. A dynamical system $\Sigma = (X, \sigma)$ is said to be equicontinuous if the collection of homeomorphisms $\{\sigma^n | n \in Z\}$ are equicontinuous.

Namely equicontinuity means that for any index α there exists an index β such that

$$(x, y) \in \beta \text{ implies } (\sigma^n x, \sigma^n y) \in \alpha \text{ for every } n \in Z \quad .$$

We shall discuss in Chap. 5 many equicontinuous dynamical systems. Hence we notice here that the dynamical system of Bernoulli shift (Example 1.1.4) is not equicontinuous. This will be seen by finding a pair of elements (x, y) of $X(k)$ with $d(x, y) = \frac{1}{1+n}$ arbitrary small but with $d(\sigma^{-n}(x), \sigma^{-n}(y)) = 1$ for some positive integer n.

Theorem 1.2.1. A minimal equicontinuous dynamical system $\Sigma = (X, \sigma)$ is strictly ergodic.

Proof. It suffices to show that Σ is uniquely ergodic. Let f be a continuous function on X. The averaging functions,

$$\{f_n(x) = \frac{1}{n} \sum_{k=0}^{n-1} \alpha^k(f)(x) \mid n = 1,2,3,\dots\}$$

are then equicontinuous. In fact, since f is uniformly continuous, for a positive number ε there exists an index α in $\mathcal{U}$ such that

$$|f(x) - f(y)| < \varepsilon \qquad \text{for every pair} \qquad (x, y) \in \alpha \qquad .$$

On the other hand, by the assumption there also exists an index β such that

$$(\sigma^n x, \sigma^n y) \in \alpha \qquad \text{for every} \qquad (x, y) \in \beta \quad , \qquad n \in Z \qquad .$$

Hence if $(x, y) \in \beta$ we have

$$|f_n(x) - f_n(y)| \le \frac{1}{n} \sum_{k=0}^{n-1} |f(\sigma^{-k}(x)) - f(\sigma^{-k}(y))| < \varepsilon \qquad ,$$

which shows the equicontinuity of the family $\{f_n\}$.

Now by the theorem of Ascoli-Arzela there exists a subsequence $\{f_{n_k}\}$ converging to a function in norm, say to g. Here the equalities

$$f_n(\sigma^{-1}x) = \frac{1}{n} \sum_{k=0}^{n-1} f(\sigma^{-(k+1)}(x))$$

$$= f_n(x) + \frac{1}{n}(f(\sigma^{-n}(x)) - f(x))$$

imply that the function $f_{n_k}(\sigma^{-1}x) - f_{n_k}(x)$ converges to 0. Hence $\alpha(g) = g$, and since Σ is minimal the function g must be constant, say c_f. Let μ be an invariant probability measure, then

$$\mu(f) = \mu(f_{n_k}) \to \mu(g) = c_f \quad .$$

Namely, $\mu(f) = c_f$ but the constant c_f does not depend on the measure μ. Therefore μ is a unique invariant measure. This completes the proof.

In general an equicontinuous system is neither necessarily uniquely ergodic nor minimal.

Definition 1.2.2. A dynamical system $\Sigma = (X, \sigma)$ is said to be distal if for any pair of points of X, (x, y), with $x \neq y$ we have

$$\overline{\{(\sigma^n x, \sigma^n y) | n \in Z\}} \cap \Delta = \phi \quad ,$$

where the upper bar means the closure.

In other words, Σ is distal if there exists a net σ^{n_ν} for a pair (x, y) such that $\sigma^{n_\nu}(x) \to z$ and $\sigma^{n_\nu}(y) \to z$ we can conclude that $x = y$.

One may easily verify that when X is a metric space with metric d, the system Σ is distal if and only if for every pair (x, y) with $x \neq y$ there exists a positive number ε such that $d(\sigma^n x, \sigma^n y) \geq \varepsilon$ for every $n \in Z$. Thus, obviously if σ preserves the metric d the system is distal but the converse is not true in general.

Proposition 1.2.2. An equicontinuous dynamical system is distal.

Proof. Take a pair of different points (x, y), then there exists an index α in $\mathcal{U}$ such that $(x, y) \notin \alpha$. On the other hand, by the equicontinuity of $\{\sigma^n\}$ there exists an index β such that

$$(u, v) \in \beta \Rightarrow (\sigma^n u, \sigma^n v) \in \alpha \quad \text{for every integer } n \quad .$$

Thus, $(\sigma^k x, \sigma^k y)$ does not belong to β for every integer k, and

$$\{(\sigma^n x, \sigma^n y) | n \in Z\} \cap \Delta \subset \{(\sigma^n x, \sigma^n y) | n \in Z\} \cap \beta = \phi \quad .$$

Since the index β is an open subset of $X \times X$ we have

$$\overline{\{(\sigma^n x, \sigma^n y) | n \in Z\}} \cap \beta = \phi \quad ,$$

and as β contains the set Δ the system Σ is distal.

The converse of this proposition is not true as shown in the next

Example 1.2.1. Let σ_θ be an irrational rotation for an irrational number $\theta \in T$. The map σ on the space $T^2 = T \times T$ (with the metric $d((t_1, s_1), (t_2, s_2)) = |t_1 - t_2| + |s_1 - s_2|$) defined by $\sigma(t, s) = (\sigma_\theta t, t + s)$ is the simplest example of the so-called Anzai skew product transformations. Transformations of this type are extensively studied both as classical and abstract dynamical systems. By the argument in Example 1.1.1 it is not so hard to see that the system (T^2, σ) is minimal. We assert that it is distal but not equicontinuous. Indeed, since $\sigma^n(t, s) = (t + n\theta, s + nt + (n-1)\theta)$ for two different points $x = (t_1, s_1)$ and $y = (t_2, s_2)$ we see that

$$d(\sigma^n x, \sigma^n y) \geq |t_1 - t_2| > 0 \quad \text{for every } n \text{ if } t_1 \neq t_2$$

and

$$d(\sigma^n x, \sigma^n y) \geq |s_1 - s_2| > 0 \quad \text{for every } n \text{ if } t_1 = t_2 \quad .$$

Thus, as mentioned before for a metric space the system Σ is distal.

Next put $t_n = t + \frac{1}{2n}$, then (t_n, s) apparently converges to (t, s) whereas

$$d(\sigma^n(t_n, s), \sigma^n(t, s)) \geq |s + nt_n - (s + nt)| = \frac{1}{2} \quad .$$

Therefore Σ is not equicontinuous.

Let X^X be the set of all maps in X with the pointwise convergence topology. By Tychonoff's theorem the space becomes a compact Hausdorff space with the semigroup structure. For a dynamical system $\Sigma = (X, \sigma)$ define the sets;

$$T(\Sigma) = \{\sigma^n | n = 0,1,2,3,\ldots\} \quad ,$$

$$E(\Sigma) = \text{closure of } T(\Sigma) \text{ in } X^X \quad ,$$

and

$$E_0(\Sigma) = \{\rho \in E(\Sigma) | \rho \text{ is continuous}\}$$

We have the inclusions, $T(\Sigma) \subset E_0(\Sigma) \subset E(\Sigma)$.

The set $E(\Sigma)$ becomes a commutative semigroup called the Ellis semigroup of Σ as well as the set $E_0(\Sigma)$. This follows from the following easily verified observation.

Lemma 1.2.3. The right multiplication in $E(\Sigma)$ is continuous as well as the left multiplication in $E_0(\Sigma)$.

In the following we shall study the structure of the semigroup $E(\Sigma)$ in connection with the base space X.

Theorem 1.2.4. The semigroup $E(\Sigma)$ becomes a group if and only if Σ is distal.

Proof. Suppose that $E(\Sigma)$ is a group and consider points x, y and z in X such that both nets $\{\sigma^{n_\nu}(x)\}$ and $\{\sigma^{n_\nu}(y)\}$ converge to z. Since $E(\Sigma)$ is a compact space by passing to a subnet we may assume that the net $\{\sigma^{n_\nu}\}$ converges to a map g in $E(\Sigma)$. Then

$$g(x) = \lim_\nu \sigma^{n_\nu}(x) = x$$
$$= \lim_\nu \sigma^{n_\nu}(y) = g(y) \quad ,$$

and as g is invertible we have that $x = y$. That is, Σ is distal. Conversely suppose that Σ is distal. It follows that each map g in $E(\Sigma)$ is one-to-one and $E(\Sigma)$ has a cancellation law; $gg_1 = gg_2 \Rightarrow g_1 = g_2$. Therefore the only idempotent map in $E(\Sigma)$ is the identity. Now take a map h and put $E_1 = \{gh | g \in E(\Sigma)\}$. We consider the collection,

$$\Phi = \{S | \text{closed nonempty subset of } E_1, S^2 \subset S\} \quad .$$

Obviously $E_1 \in \Phi$ and Φ is not empty. Assume the order in Φ by inclusion, then as in the proof of (1.1.2) we may apply Zorn's lemma to the family Φ. Let S_0 be a minimal set in Φ. Take an element g in S_0 and write

$$S_0 g = \{fg \mid f \in S_0\} \quad .$$

Since $S_0^2 \subset S_0$, $S_0 g$ is a subset of S_0 and moreover $(S_0 g)^2 \subset S_0 g$. Hence $S_0 g = S_0$, which means that there is an element $f \in S_0$ such that $fg = g$. Therefore, the set

$$W = \{k \in S_0 \mid kg = g\}$$

is not empty. It is furthermore closed and satisfies the condition, $W^2 \subset W$. Therefore, $W = S_0$ and $g \in W$. This means that $g^2 = g$ and as mentioned before the identity map belongs to E_1. After all, the map h has a left inverse in $E(\Sigma)$. Since h was arbitrary, $E(\Sigma)$ is a group.

Let $\Sigma = (X, \sigma)$ be a topological dynamical system and suppose that X is expressed as a union of disjoint invariant closed sets $\{X_j\}_{j \in J}$. In this case, putting $\sigma_j = \sigma | X_j$ we say that Σ is the direct sum of the dynamical systems $\{\Sigma_j = (X_j, \sigma_j)\}_{j \in J}$ and write $\Sigma = \bigoplus_J \Sigma_j$. A dynamical system expressed as a direct sum of minimal dynamical systems is said to be semisimple. With this definition we have

<u>Proposition 1.2.5.</u> Every distal system is semisimple.

<u>Proof.</u> Suppose Σ is distal and take $x \in X$. Let $y \in \overline{O(x)}$, then we have a net $\{\sigma^{n_\nu}(x)\}$ converging to y. By passing to a subnet, we may assume that the net $\{\sigma^{n_\nu}\}$ converges to a map g in $E(\Sigma)$ and $g(x) = y$. Since $E(\Sigma)$ is a group we have $x = g^{-1}(y)$ with $g^{-1} \in E(\Sigma)$. We then see that $x \in \overline{O(y)}$ and $\overline{O(x)} = \overline{O(y)}$. Hence, $\overline{O(x)}$ is a minimal set and X is a disjoint union of minimal sets for σ. Namely, Σ is semisimple.

<u>Theorem 1.2.5.</u> If a dynamical system $\Sigma = (X, \sigma)$ is equicontinuous, then the semigroup $E(\Sigma)$ becomes an abelian topological group of homeomorphisms of X, whence $E(\Sigma) = E_0(\Sigma)$.

<u>Proof.</u> By (1.2.2) and the above theorem $E(\Sigma)$ is an abelian group. We assert that every map g in $E(\Sigma)$ is continuous so that g becomes a homeomorphism. In fact, take a point x_0 and fix. Let $\{\sigma^{n_\nu}\}$ be a net converging to g. Take an arbitrary index α and choose indexes α_1 and α_2 such that $\alpha_2^2 \subset \alpha_1$ and $\overline{\alpha_1(g(x))} \subset \alpha(g(x))$. Since $\{\sigma^n\}$ is equicontinuous, there exists an index β such that

$$(x, y) \in \beta \Rightarrow (\sigma^n x, \sigma^n y) \in \alpha_2 \quad \text{for every } n.$$

On the other hand, for the net $\{\sigma^{n_\nu}(x)\}$ there is an integer N_0 such that $\sigma^{n_\nu}(x) \in \alpha_2(g(x))$ for every $n_\nu \geq N_0$ (with respect to the net order). Now if $y \in \beta(x)$ then

$$(\sigma^{n_\nu}(x), \sigma^{n_\nu}(y)) \in \alpha_2 \quad .$$

Hence, $(g(x), \sigma^{n_\nu}(y)) \in \alpha_1$, that is, $\sigma^{n_\nu}(y) \in \alpha_1(g(x))$. From the choice of α_1, we see then that $g(y) \in \alpha(g(x))$. Thus, g is continuous at the point x.

In order to show that $E(\Sigma)$ becomes even a topological group we need, however, comparatively long discussions which are too long to be included here. The difficult part is to show that multiplication in $E(\Sigma)$ is jointly continuous under the observation Lemma 1.2.3. Once we have established this fact the rest of the proof is rather easy to derive. Indeed, identifying each element $g \in E(\Sigma)$ with the map $\hat{g}$; $h \to gh$ in $E(\Sigma)$ one may see that the pointwise convergence topology of the set $\{\hat{g} | g \in E(\Sigma)\}$ which coincides with the original topology of $E(\Sigma)$ turns out to be equivalent to the uniform convergence topology on $E(\Sigma)$ because $E(\Sigma)$ is a compact space. Then the latter topology in $\{\hat{g} | g \in E(\Sigma)\}$ is known to be compatible with the group structure of $E(\Sigma)$. We leave the difficult part to the original paper, Ellis[8].

<u>Theorem 1.2.6.</u> Suppose a dynamical system $\Sigma = (X, \sigma)$ is equicontinuous and topologically transitive, then the topological group $E(\Sigma)$ is homeomorphic to X.

Thus in this case the space X may be regarded as a compact abelian group.

Proof. We note first that the system Σ is minimal, for by (1.2.2) and (1.2.5) Σ becomes semisimple. Take a point x_0 and fix Define the map

$$\Phi \; ; \quad g \quad E(\Sigma) \to g(x_0) \in X \qquad .$$

We assert that this map is a homeomorphism between $E(\Sigma)$ and X. The map is apparently continuous and moreover one-to-one. Indeed, if $g(x_0) = h(x_0)$ we have that $g(\sigma^n(x_0)) = h(\sigma^n(x_0))$ $(n = 0,1,2,\dots)$ because both maps g and h commutes with those maps σ^n's by definition, whence $g = h$ since the positive orbit $O_\sigma^+(x_0)$ is dense in X by (1.1.1-(2)). Furthermore the map is surjective for the image $\Phi(E(\Sigma))$ is compact and contains the set $O_\sigma^+(x_0)$, hence $\Phi(E(\Sigma)) = X$. Therefore, the inverse map Φ^{-1} from X to $E(\Sigma)$ is continuous, too.

APPENDIX A

MONOTHETIC GROUP

A topological group G is said to be monothetic if it has a dense subgroup $\{x_n | n \in Z\}$ which is a homomorphic image of Z. The group is characterized in the following:

Theorem A. Let G be an abelian topological group.

(1) If G is not compact, G is monothetic if and only if $G = Z$;

(2) If G is compact, G is monothetic if and only if the dual group Γ is a subgroup of T_d, the torus group with the discrete topology.

Proof. If G is discrete, then either $G = Z$ or G is a finite cyclic group. Hence it is compact if $G \neq Z$. Thus, for the first assertion it is enough to show that G is compact if G is not discrete. Let U be a symmetric neighborhood of 0 in G with compact closure $\bar{U}$. Take an element y in G, then $y \in x_k + U$ for some x_k and there is a symmetric neighborhood V of 0 in G such that $y - x_k + V \subset U$. Since G is not discrete, the neighborhood V contains infinitely many points x_n, and since V is symmetric, $x_{-n} \in V$ if $x_n \in V$. Hence there exists an integer $j < k$ with $x_j \in V$. Put $i = k - j$, then $i > 0$ and

$$y - x_i = y - x_k + x_j \in y - x_k + V \subset U \quad .$$

Hence $y \in x_i + U$, which shows that

$$G = \bigcup_{i=1}^{\infty} (x_i + U) \quad .$$

On the other hand, since $\bar{U}$ is compact, it follows that

$$\bar{U} \subset \bigcup_{i=1}^{N} (x_i + U) \quad \text{for some integer } N \quad .$$

Now for every $y \in G$ let $n = n(y)$ be the smallest positive integer such that $y \in x_n + \bar{U}$. Then $x_n - y \in x_i + U$ for some i $(1 \le i \le N)$, and $y \in x_{n-i} + U$. Since $i > 0$, $n - i < n$ whence $n - i \le 0$ from our choice of n. Thus $n \le i \le N$ for all y and we have

$$G = \bigcup_{i=1}^{N} (x_i + \bar{U}) \quad .$$

Therefore, G is compact. This completes the proof.

As for the assertion (2) we observe that if G is monothetic the continuous characters of G are evidently determined by their values on the dense homomorphic image of Z in G. It follows that the dual group Γ is a subgroup of the dual T of Z. Since G is compact, Γ must be discrete.

Conversely if the dual group Γ of G is a subgroup of T_d, then by the duality theorem G is a quotient subgroup of the dual group of T_d, which is known to be the Bohr compactification $\hat{Z}$ of Z. Since $\hat{Z}$ is clearly a monothetic group, its continuous homomorphic image is again monothetic.

CHAPTER 2
ELEMENTARY THEORY OF C*-ALGEBRAS

In this chapter, together with the first half of the next chapter we prepare the results from the theory of C*-algebras for our subsequent discussion. It is however far beyond the size of the present text to give full details for all results introduced in these two chapters even if we restrict to the proofs of important theorems among them. Thus these parts are designed so as to give the readers an understanding of the concept of C*-algebras with related things and to provide the results which will be used later with or without their proofs. Readers who are interested in further details may consult those books such as [C], [D] and [E].

2.1 The Gelfand Representation of Commutative Banach Algebras and Spectral Calculus in Banach Algebras

In order to understand how we treat the algebra $C(X)$ or C*-algebras we introduce first the concept of Banach algebras.

Definition 2.1.1. A Banach algebra is an algebra (over the complex numbers $\mathbb{C}$ in general) whose linear structure forms a Banach space and the multiplication satisfies the condition;

$$\|xy\| \le \|x\| \, \|y\| \quad .$$

By an involution or *-operation of a Banach algebra we mean an isometric conjugate linear anti-automorphism $x \to x^*$ such that

$x^{**} = x$. A Banach algebra with an involution is called a Banach *-algebra.

When a Banach algebra A has a unit (denoted by 1_A or 1 if no confusion arises) we assume that $\|1\| = 1$. The algebra $C(X)$ is a commutative Banach algebra with unit (i.e. unital), whereas the algebra $C_0(Y)$ of continuous functions on a locally compact space Y vanishing at infinity is an example of a commutative Banach algebra without unit (i.e. nonunital) with the same supremum norm. Moreover both algebras are Banach *-algebras with the involution $f^*(x) = \overline{f(x)}$ (complex conjugate). Other examples of Banach algebras are such as the algebra $L^\infty(\Omega, \mu)$ on a probability measure space $\{\Omega, \mathscr{B}, \mu\}$ with the essential supremum norm and the convolution algebra $L^1(G)$ for a locally compact (unimodular) group G with the Haar measure ds etc.

When a Banach algebra A is not unital, there is a way to adjoin a unit to A. Namely, we consider the set

$$A_1 = \{(x, \lambda) \mid x \in A, \ \lambda \in \mathbb{C}\}$$

together with natural addition, scalar multiplication and the multiplication,

$$(x, \lambda)(y, \mu) = (xy + \mu x + \lambda y, \ \lambda\mu)$$

and the norm

$$\|(x, \lambda)\| = \|x\| + |\lambda| \quad .$$

The algebra A_1 then becomes a Banach algebra with unit $(0, 1)$. Thus, identifying $(x, 0)$ with x and $(0, \lambda)$ with λ an element (x, λ) in A_1 is written as $(x, \lambda) = x + \lambda$. Here, A is the largest ideal of A_1.

Henceforth in this section we mainly assume that the Banach algebra A has a unit 1.

Theorem 2.1.1. Suppose that an element a of A satisfies the condition, $\|1 - a\| < 1$, then a is invertible. The set of all invertible elements, G, forms a topological group which is open in A.

Proof. From the assumption

$$\sum_{n=0}^{\infty} \|(1-a)^n\| < \sum_{n=0}^{\infty} \|1-a\|^n < \infty \quad ,$$

hence the series $\sum_{n=0}^{\infty} (1-a)^n$ is summable, say to $b \in A$. Then,

$$ba = b(1-(1-a)) = b - \sum_{n=1}^{\infty} (1-a)^n = 1$$

and similarly $ab = 1$, i.e. $b = a^{-1}$.

Next take an element a of G, then the neighborhood of a, $\{b \mid \|a-b\| < 1/\|a^{-1}\|\}$ is contained in G. In fact, since

$$\|1-a^{-1}b\| = \|a^{-1}(1-b)\| \le \|a^{-1}\| \, \|1-b\| < 1 \quad ,$$

we have that $a^{-1}b \in G$ and $b \in G$. Thus G is open. In order to see that G is a topological group it is enough to show that the inverse operation is continuous. Now suppose that $\|1-a\| < \varepsilon$, then

$$\|a^{-1}\| \le \sum_{n=0}^{\infty} \|1-a\|^n < \frac{1}{1-\varepsilon} \quad .$$

Hence,

$$\|1-a^{-1}\| \le \|a^{-1}\| \, \|a-1\| < \frac{1}{1-\varepsilon} \quad ,$$

which converges to zero as $\varepsilon \to 0$. Therefore the operation is continuous at the unit and then we obtain the following series of implications,

$$a_n \to a \quad \text{in} \quad G \Rightarrow a^{-1}a_n \to 1$$

$$\Rightarrow a_n^{-1}a \to 1 \Rightarrow a_n^{-1} \to a^{-1} \quad .$$

This shows the continuity of the inverse operation.

Let I be a closed ideal of A, then the quotient algebra A/I is a Banach space and one may further verify that it is a Banach algebra. On the other hand, for a proper ideal I a usual argument by Zorn's lemma shows the existence of a maximal ideal M containing I. Since there exists no dense proper ideal in a unital Banach algebra by the first part of the above theorem, M is necessarily closed. This fact also implies that for a unital Banach algebra the algebraic simplicity is equivalent to the topological simplicity for closed ideals. Besides, the same reason shows the following easily verified fact.

Proposition 2.1.2. Suppose that A is commutative, then an element x is invertible if and only if there is no maximal ideal containing x.

The next theorem plays a key rôle in the Gelfand representation theory.

Theorem 2.1.3. Let A be a commutative Banach algebra and M be a maximal ideal. Then the quotient algebra A/M is isometric isomorphic to the complex number field $\mathbb{C}$.

Proof. Any nonzero element a of A/M is invertible by the above proposition, hence the algebra A/M is a field. We assert that there is a complex number λ_a such that $a = \lambda_a \hat{1}$ where $\hat{1}$ means the class for 1. Thus contrary, suppose that $a - \lambda\hat{1} \neq 0$ for every complex number λ. Let φ be a bounded linear functional of A/M and set $f(\lambda) = \varphi((a - \lambda\hat{1})^{-1})$. The function $f(\lambda)$ is then analytic on the whole plane $\mathbb{C}$. Indeed, $f'(\lambda) = \varphi((a - \lambda\hat{1})^{-2})$ and moreover it is bounded because $f(\lambda) \to 0$ as $\lambda \to \infty$. Therefore by Liouville's theorem $f(\lambda)$ is constant and $f(\lambda) = 0$. Since φ is arbitrary, this means that $(a - \lambda\hat{1})^{-1} = 0$, a contradiction. It follows that there exists a constant λ_a such that $a = \lambda_a \hat{1}$, and the map $a \to \lambda_a$ is easily seen to be an isometric isomorphism between A/M and $\mathbb{C}$. This completes the proof.

We call a multiplicative functional φ on A a character. If φ is nonzero the kernel $M_\varphi = \varphi^{-1}(0)$ is a maximal ideal and closed.

The map φ is a composition of the quotient homomorphism of A to A/M and the isomorphism $\hat{a} = \varphi(a)\hat{1} \to \varphi(a)$ of A/M to $\mathbb{C}$. Hence a character φ is necessarily bounded with $\|\varphi\| = 1$. Conversely the above theorem shows that a maximal ideal M of A defines a character φ_M with $\varphi_M^{-1}(0) = M$. Since the set $\Delta(A)$ of all (nonzero) characters on A is in the unit sphere of A^*, the dual of A, it is easily seen to be compact in the weak *-topology.

Now let $\mathcal{M}(A)$ be the set of all maximal ideals of A. We identify $\mathcal{M}(A)$ with the set $\Delta(A)$ by the above correspondence together with the topology. Then each element a in A yields a continuous function $\hat{a}$ on the compact space $\mathcal{M}(A)$ such that $\hat{a}(M) = \varphi_M(a)$.

<u>Definition 2.1.2.</u> We call the above homomorphism

$$\Phi \ ; \ a \in A \to \hat{a}(M) \in C(\mathcal{M}(A))$$

the Gelfand representation of A.

Let us consider the simplest case $A = C(X)$. In this case, a character on $C(X)$ necessarily becomes an evaluation functional φ_x at some point x. In fact, let M be a maximal ideal of $C(X)$ and assume, on the contrary, that there is no point on which every function of M vanishes. Then, at each point x we can find a function f_x of M and a neighborhood U_x of x such that $f_x(y) \neq 0$ for $y \in U_x$. Since X is compact, one may find a finite covering $\{U_{x_i}\}_{i=1}^{n}$ and we can define a function $f(x) = \sum_{i=1}^{n} f_{x_i}(x)\overline{f_{x_i}(x)}$ of M. By definition, $f(x) > 0$ for every $x \in X$ and f is invertible, a contradiction. Let $M = \varphi_{x_0}^{-1}(0)$ for some point x_0. By the correspondence $x_0 \to \varphi_M = \varphi_{x_0}$ the space X becomes homeomorphic to the character space $\Delta(C(X))$. Therefore, identifying these two spaces the identity map on $C(X)$ may be regarded as the Gelfand representation of $C(X)$.

When the algebra A is not unital the space $\Delta(A)$ forms a locally compact space in the weak *-topology and $\mathcal{M}(A)$ consists of all regular maximal ideals of A. The Gelfand representation is a homomorphism of A into the algebra $C_0(\mathcal{M}(A))$, whereas the space $\Delta(A_1)$ of A_1 is a one-point compactification of $\Delta(A)$ in which the ideal point of the infinity corresponds to the character φ_0 on A_1 such that $\varphi_0(A) = 0$.

In general the map is neither onto nor one-to-one. If it is one-to-one A is said to be semisimple.

Definition 2.1.3. The spectrum $sp(a)$ of an element a of a unital Banach algebra A is the set;

$$sp(a) = \{\lambda \in \mathbb{C} | a - \lambda \text{ is not invertible}\} \quad .$$

The complement of $sp(a)$ in $\mathbb{C}$ is called the resolvent of a.

Since the map $\lambda \in \mathbb{C} \to a - \lambda \in A$ is continuous, the second half of (2.1.1) shows that the resolvent set is open. Hence the set $sp(a)$ is closed.

If A is not unital we denote $sp(a)$ the spectrum of a in the algebra A_1. In this case, $sp(a)$ always contains zero.

Lemma 2.1.4. If the algebra A is unital,

$$sp(ab) \cup \{0\} = ap(ba) \cup \{0\}$$

for any $a, b \in A$.

Proof. For a complex number $\lambda \notin sp(ab) \cup \{0\}$, we assert that $ba - \lambda$ is invertible, i.e. $\lambda \notin sp(ba) \cup \{0\}$. This will prove the fact by symmetry. Let u be the inverse of $ab - \lambda$, then

$$abu = uab = 1 + \lambda u \quad .$$

Hence,

$$(ba - \lambda)(bua - 1) = \lambda = (bua - 1)(ba - \lambda)$$

and $ba - \lambda$ is invertible.

Now define the spectral radius of a in A by

$$\|a\|_{sp} = \sup\{|\lambda| \mid \lambda \in sp(a)\} .$$

Theorem 2.1.5. The spectrum $sp(a)$ of a in a Banach algebra A is a nonempty compact set and

$$\|a\|_{sp} = \lim_{n\to\infty} \|a^n\|^{\frac{1}{n}} . \tag{2.1}$$

Proof. Suppose that $sp(a) = \phi$ for an element a, which means that $a - \lambda$ is invertible for every complex number λ. Then the same argument as in the proof of (2.1.3) leads us to a contradiction. Hence, $sp(a)$ is a nonempty closed subset of $\mathbb{C}$. We assert that $sp(a)$ is bounded. Indeed, if $|\lambda| > \|a\|$, then $1 - \frac{1}{\lambda}a$ is invertible by (2.1.1). Hence $\lambda - a$ is invertible and $\lambda \notin sp(a)$. Thus $sp(a)$ is contained in the disk $|\lambda| \leq \|a\|$.

Next, take $\lambda \in sp(a)$. Then one may easily verify that $\lambda^n \in sp(a^n)$. It follows that

$$\|a\|_{sp} \leq (\|a^n\|_{sp})^{\frac{1}{n}} \leq \|a^n\|^{\frac{1}{n}}$$

and

$$\|a\|_{sp} \leq \liminf \|a^n\|^{\frac{1}{n}} .$$

Let φ be a bounded linear functional on A. We see then that the function $f(\lambda) = \varphi((a-\lambda)^{-1})$ is holomorphic for $|\lambda| > \|a\|_{sp}$. Therefore its Laurent expansion $\sum_{n=0}^{\infty} \frac{\varphi(a^n)}{\lambda^{n+1}}$ converges in the same domain. It follows that

$$\lim_{n\to\infty} \frac{\varphi(a^n)}{\lambda^{n+1}} = 0 \qquad \text{for every} \qquad \varphi \in A^* .$$

Therefore, by the uniform boundedness principle, the sequence $\frac{1}{\lambda^{n+1}} a^n$ is norm bounded. Thus for each such λ there exists a number $B > 0$ such that

$$\left\| \frac{1}{\lambda^{n+1}} a^n \right\| \leq B \quad , \qquad \text{i.e.} \qquad \| a^n \| \leq B|\lambda|^{n+1} \quad .$$

Hence,

$$\limsup \| a^n \|^{\frac{1}{n}} \leq \lim_{n\to\infty} B^{\frac{1}{n}} |\lambda| = |\lambda| \quad ,$$

and

$$\limsup \|a^n\|^{\frac{1}{n}} \leq \|a\|_{sp} \quad .$$

This completes the proof.

If moreover A is commutative the fact (2.1.2) shows that

$$sp(a) = \{\hat{a}(M) | M \in \mathscr{M}(A)\}$$

and

$$\| a \|_{sp} = \| \hat{a} \|_{\infty} \quad .$$

Let a be an element of a Banach algebra A and f be a holomorphic function in a neighborhood U_f of $sp(a)$. Let C be a smooth simple closed curve enclosing $sp(a)$. We consider an A-valued continuous function: $\lambda \to \frac{1}{2\pi i} f(\lambda)(\lambda - a)^{-1}$ on C. Then as in the case of the theory of complex functions one can define the integration along the curve C and show the existence of this integration. Put

$$f(a) = \frac{1}{2\pi i} \int_C f(\lambda)(\lambda - a)^{-1} d\lambda \quad .$$

Note that for a functional $\varphi \in A^*$ we have

$$< f(a), \varphi > = \frac{1}{2\pi i} \int_C f(\lambda) < (\lambda - a)^{-1}, \varphi > d\lambda \quad .$$

Thus by Cauchy's theorem we see that the element $f(a)$ does not depend on the choice of the curve C. With this definition we have the following

<u>Proposition 2.1.6.</u> The map: $f \to f(a)$ is a homomorphism of the algebra of all functions holomorphic in a neighborhood of $sp(a)$ into A, which send the constant function 1 to the identity 1 of A and the function $f(\lambda) = \lambda$ to a.

<u>Proof.</u> It is enough to show that the map is multiplicative. Let f and g be two functions holomorphic in neighborhoods U_f and U_g of $sp(a)$. Put $U = U_f \cap U_g$ and let C_1 and C_2 be simple smooth closed curves in U enclosing $sp(a)$ such that C_2 lies inside the curve C_1. Then we have

$$f(a)g(a) = \left(\frac{1}{2\pi i}\int_{C_1} f(\lambda)(\lambda - a)^{-1}d\lambda\right)\left(\frac{1}{2\pi i}\int_{C_2} g(\mu)(\mu - a)^{-1}d\mu\right)$$

$$= -\frac{1}{4\pi^2}\iint_{C_1 \times C_2} f(\lambda)g(\mu)(\lambda - a)^{-1}(\mu - a)^{-1}d\lambda d\mu$$

$$= -\frac{1}{4\pi^2}\iint_{C_1 \times C_2} f(\lambda)g(\mu)\frac{1}{\lambda - \mu}\{(\lambda - a)^{-1} - (\mu - a)^{-1}\}d\lambda d\mu$$

$$= -\frac{1}{4\pi^2}\iint_{C_1 \times C_2} \frac{f(\lambda)g(\mu)}{\lambda - \mu}(\mu - a)^{-1}d\lambda d\mu$$

$$+ \frac{1}{4\pi^2}\iint_{C_1 \times C_2} \frac{f(\lambda)g(\mu)}{\lambda - \mu}(\lambda - a)^{-1}d\lambda d\mu \quad .$$

Here the second term is equal to

$$\frac{1}{4\pi^2}\int_{C_1}\left(\int_{C_2}\frac{g(\mu)}{\lambda - \mu}d\mu\right)f(\lambda)(\lambda - a)^{-1}d\lambda = 0$$

because the function $\frac{g(\mu)}{\lambda - \mu}$ is holomorphic inside the curve C_1 if λ lies on C_1. Hence

$$f(a)g(a) = \frac{1}{2\pi i}\int_{C_2}\left(\frac{1}{2\pi i}\int_{C_1}\frac{f(\lambda)}{\lambda-\mu}\,d\lambda\right)g(\mu)(\mu-a)^{-1}d\mu$$

$$= \frac{1}{2\pi i}\int_{C_2} f(\mu)g(\mu)(\mu-a)^{-1}d\mu$$

$$= (fg)(a) \quad .$$

Next suppose $f(\lambda)\equiv 1$, then

$$f(a) = \frac{1}{2\pi i}\int_C (\lambda-a)^{-1}d\lambda \quad .$$

Here one may choose a circle as the curve C whose radius is greater than $\|a\|$. We have then

$$(\lambda-a)^{-1} = \sum_{n=0}^{\infty}\frac{a^n}{\lambda^{n+1}}$$

uniformly for every $\lambda\in C$. Hence we get

$$f(a) = \frac{1}{2\pi i}\int_C \sum_{n=0}^{\infty}\frac{a^n}{\lambda^{n+1}}\,d\lambda$$

$$= \frac{1}{2\pi i}\sum_{n=0}^{\infty} a^n \int_C \frac{1}{\lambda^{n+1}}\,d\lambda = 1 \quad .$$

Similarly $f(a)=a$ if $f(\lambda)=\lambda$.

<u>Theorem 2.1.7.</u> (Spectral mapping theorem). Let a be an element of a Banach algebra A. If f is a holomorphic function in a neighborhood of $sp(a)$, then

$$sp(f(a)) = f(sp(a)) \quad .$$

Furthermore if g is a holomorphic function in a neighborhood of $f(sp(a))$, we have

$$g\cdot f(a) = g(f(a)) \quad .$$

Proof. Take a number $\mu \notin f(sp(a))$. Then the function $h(\lambda) = \frac{1}{f(\lambda) - \mu}$ is holomorphic in a neighborhood of $sp(a)$. By the above proposition $h(a)$ is the inverse of $f(a) - \mu$ and $\mu \notin sp(f(a))$, that is, $sp(f(a)) \subset f(sp(a))$. Let $\mu \in f(sp(a))$, then $\mu = f(\lambda_0)$ for some $\lambda_0 \in sp(a)$. Therefore, there exists a holomorphic function k in a neighborhood of $sp(a)$ such that

$$f(\lambda) - \mu = (\lambda - \lambda_0)k(\lambda) \quad .$$

Hence by the above proposition,

$$f(a) - \mu = (a - \lambda_0)k(a) = k(a)(a - \lambda_0) \quad .$$

This means that $f(a) - \mu$ is not invertible and $\mu \in sp(f(a))$. Thus, $f(sp(a)) = sp(f(a))$.

Next choose smooth simple closed curves C_1 and C_2 in such a way that C_2 encloses $sp(a)$ in the domain of f and C_1 encloses $f(sp(a))$ together with the image of C_2 by f in the domain of g. We have

$$\begin{aligned} g \cdot f(a) &= \frac{1}{2\pi i} \int_{C_1} g(f(\lambda))(\lambda - a)^{-1} d\lambda \\ &= -\frac{1}{4\pi^2} \int_{C_1} \left(\int_{C_2} \frac{g(\mu)}{\mu - f(\lambda)} d\mu \right) (\lambda - a)^{-1} d\lambda \\ &= -\frac{1}{4\pi^2} \int_{C_2} g(\mu) \left\{ \int_{C_1} (\mu - f(\lambda))^{-1} (\lambda - a)^{-1} d\lambda \right\} d\mu \\ &= \frac{1}{2\pi i} \int_{C_2} g(\mu)(\mu - f(a))^{-1} d\mu \\ &= g(f(a)) \quad . \end{aligned}$$

Now the exponential function $\exp \lambda$ is an entire function so that we can define the element $\exp a$ for every $a \in A$. This element is also defined by the norm convergent power series;

$$\exp a = \sum_{n=0}^{\infty} \frac{1}{n!} a^n \quad .$$

If a and b are commuting one can apply the same calculus as in scalar case and obtain the formula,

$$\exp(a+b) = \exp a \exp b \quad .$$

The logarithm of an element a is not always defined because $\log \lambda$ is not an entire function. If $sp(a)$ is contained in the domain of the principal logarithm $\operatorname{Log} \lambda$, then $\log a$ is defined as the element $\operatorname{Log} a$. By the above result

$$\exp \log a = a$$

for such an element a. Also applying the same result we have the assertion that if $sp(a)$ is contained in the open strip: $-\pi < \operatorname{Im}\lambda < \pi$ then $\log \exp a$ is defined and is equal to a.

2.2 Elementary Properties of C*-algebras

Definition 2.2.1. A C*-algebra A is an involutive Banach algebra whose involution satisfies the relation,

$$\| a^*a \| = \| a \|^2 \quad . \tag{2.2}$$

An element a of A is said to be self-adjoint if $a = a^*$, normal if $a^*a = aa^*$, unitary if $a^*a = aa^* = 1$ when A is unital, projection if $a = a^*$ and $a^2 = a$. A subset of A is said to be self-adjoint if $a \in S$ implies $a^* \in S$. We write A_h as the real Banach space of all self-adjoint elements of A.

Let H be a Hilbert space and $B(H)$ be the algebra of all bounded linear operators on H. Then the *-operation in $B(H)$ satisfies the condition (2.2), so that any uniformly closed self-adjoint subalgebra of $B(H)$ provides an example of C*-algebras. The algebra of all compact operators on H, C(H), is an example of non-unital C*-algebras if H is infinite dimensional. These algebras are naturally called concrete C*-algebras comparing with abstract C*-algebras defined above. Those kind of elements in C*-algebras such as self-adjoint elements, normal ones, are named after corresponding operators in concrete C*-algebras.

There were considerable materials for discussions about representations of an abstract C*-algebra as a concrete C*-algebra as well as about the axiom (2.2) until the present theory of C*-algebras has been established. In case of commutative C*-algebras, there is also another problem of the Gelfand representation. However, first of all we must discuss the adjunction of a unit for a non-unital C*-algebra. When a C*-algebra A is non-unital the unit adjoined Banach algebra A_1 defined before is not a C*-algebra. Therefore we must adjust the norm in A_1. For an element x of A_1 we consider a bounded linear operator L_x on A as the left multiplication. Then the axiom (2.1) shows that $\| L_a \| = \| a \|$ if $a \in A$, and it turns out that $\| x \| \equiv \| L_x \|$ for x in A_1 is indeed a C*-norm in A_1. In fact, if $L_x = 0$ for some nonzero element $x = x' + \lambda 1 (x' \in A)$ then $\lambda \neq 0$ and we have, for any $a \in A$, that

$$0 = \frac{1}{\lambda} xa = \frac{1}{\lambda} x'a + a \quad .$$

Hence, $-\frac{1}{\lambda} x'$ is a left unit of A and $(-\frac{1}{\lambda} x')^*$ is a right unit of A. Therefore, A has a unit, $-\frac{1}{\lambda} x' = (-\frac{1}{\lambda} x')^*$, a contradiction. Next, for any $x \in A_1$ and $\varepsilon > 0$ there exists an $a \in A$ with $\| a \| \leq 1$ such that

$$(1 - \varepsilon) \| x \| \leq \| xa \| \quad ,$$

which implies that

$$\begin{aligned} \| x^*x \| &\geq \| a^*(x^*x)a \| = \| (xa)^*(xa) \| \\ &= \| xa \|^2 \geq (1 - \varepsilon)^2 \| x \|^2 \quad . \end{aligned}$$

Therefore,

$$\| x\|^2 \leq \| x^*x\| \leq \| x^*\| \, \| x\| \quad , \quad \text{and} \quad \| x\| \leq \| x^*\| \quad .$$

It follows that $\| x\| = \| x^*\|$ and $\| x^*x\| = \| x\|^2$. Finally, since A is complete and of codimension one in A_1, A_1 is also complete.

We must however notice that this procedure does not always overcome the problems for non-unital C*-algebras. Sometimes lack of a unit changes situations drastically from unital case and there appear much more complicated situations in the problems about which we must directly analyse. Therefore in the theory of C*-algebras whether or not the algebra has a unit is an important assumption. But since we shall only treat discrete topological dynamical systems on compact spaces and since this means that we shall be exclusively concerned with unital C*-algebras we henceforth assume that *a C*-algebra always has a unit throughout our discussions unless otherwise stated.* Readers interested in non-unital case may consult those books referred before.

Theorem 2.2.1. Let A be a commutative C*-algebra, then the Gelfand representation of A is an isometric *-isomorphism of A onto the algebra $C(\mathscr{M}(A))$.

Here a *-isomorphism means an isomorphism preserving *-operation. Similarly we shall use the words, *-homomorphism, *-representation and so on.

The above result says that to consider a commutative C*-algebra is nothing but to handle with the algebra $C(X)$. The proof of this theorem depends on the following two basic properties of elements of C*-algebras;

For a normal element a, $\| a\| = \| a\|_{sp}$ (2.3)

$sp(h)$ R(real number) for a selfadjoint element h (2.4)

Once these assertions are proved the rest of the proof is rather easy. Namely, at first we know that the Gelfand representation map Φ is isometric by (2.3). Moreover, since an element a of A is written as $a = h_1 + ih_2$ for self-adjoint elements h_1, h_2 (real and imaginary parts of a) we have by (2.4)

$$\hat{a}^*(M) = \varphi_M(h_1 - ih_2) = \varphi_M(h_1) - i\varphi_M(h_2)$$
$$= \hat{h}_1(M) - i\hat{h}_2(M) = \overline{\hat{a}(M)} \quad .$$

Thus, the image $\Phi(A)$ is a closed self-adjoint unital subalgebra of $C(\mathscr{M}(A))$ separating the points of (A). Hence, $\Phi(A) = C(\mathscr{M}(A))$ by the Stone-Weierstrass theorem.

<u>Proof of (2.3).</u> The equalities

$$\| a^{2n}\|^2 = \| (a^{2n})^*(a^{2n}) \| = \| (a^*a)^{2n}\|$$
$$= \| (a^*a)^n\|^2$$

implies that

$$\| a^{2^n} \| = \| (a^*a)^{2^{n-1}} \| = \dots = \| a^*a\|^{2^{n-1}}$$
$$= \| a\|^{2^n} \quad .$$

Hence, by (2.1.5)

$$\| a\| = \lim_{n\to\infty} \| a^{2^n} \|^{2^{-n}} = \| a\|_{sp} \quad .$$

<u>Proof of (2.4).</u> Consider the following convergent series in A,

$$u = \sum_{n=0}^{\infty} \frac{(ih)^n}{n!} \; (= \exp(ih)) \quad .$$

This defines a unitary i.e. $uu^* = u^*u = 1$. Hence

$$1 = \|1\| = \| u^*u\| = \|u\|^2 = \| u^*\|^2 \quad ,$$

and by (2.3)

$$\|u\|_{sp} = \|u^*\|_{sp} = 1 \quad .$$

Combining these things with set equalities

$$sp(u^{-1}) = \{\lambda^{-1} | \lambda \in sp(u)\}$$

$$= \{\bar{\lambda} | \lambda \in sp(u)\} \quad ,$$

one sees that the set $sp(a)$ must be contained in the unit circle of $\mathbb{C}$. On the other hand, since $\exp i(sp(h)) = sp(\exp ih)$ by (2.1.7), this is possible only if $sp(h) \subseteq R$.

We must consider later topological dynamical system arising in the spectrums of commutative C*-algebras. In these situations discussions for separable C*-algebras correspond to usual topological dynamical systems for compact metric spaces, but even if we discuss on a separable Hilbert space, those concrete (commutative) C*-algebras are often highly non-separable.

Let A be a C*-algebra. We call a subalgebra B of A a C*-subalgebra if it is closed and self-adjoint. A unital C*-subalgebra B of A is a C*-subalgebra with the same unit of A.

Proposition 2.2.2. Let B be a unital C*-subalgebra of a C*-algebra A, then $sp_A(a) = sp_B(a)$ for every $a \in B$.

Proof. It is clear that $sp_A(a) \subset sp_B(a)$. Suppose first that a is self-adjoint and take a scalar $\lambda \notin sp_A(a)$. We assert that $\lambda \notin sp_B(a)$. Since by (2.4) $sp_B(a) \subset R$ we may assume that λ is real. Then for any $\varepsilon > 0$, $\lambda_\varepsilon = \lambda + i\varepsilon$ is not in $sp_B(a)$, so that $(a - \lambda_\varepsilon)^{-1}$ exists in B. On the other hand, $(a - \lambda)^{-1}$ exists in A, hence by (2.1.1) $(a - \lambda_\varepsilon)^{-1}$ converges to $(a - \lambda)^{-1}$ as $\varepsilon \to 0$. As B is closed, $(a - \lambda)^{-1} \in B$ and $\lambda \notin sp_B(a)$. Thus, $sp_A(a) = sp_B(a)$. Now if an element a of B is invertible in A, the self-adjoint element a^*a is invertible in A hence in B by the above argument. Therefore a is left invertible and similarly a becomes right invertible. Thus a is invertible. Applying this argument for $a-\lambda$ we complete the proof.

Let a be a normal element of A and let B be the C*-subalgebra generated by a and the unit. By (2.2.1) B is regarded as the algebra $C(\mathcal{M}(B))$ and

$$sp_A(a) = sp_B(a) = \{\hat{a}(M) | M \in \mathcal{M}(B)\} \quad .$$

The continuous function $\hat{a}$, however, separates the points of $\mathcal{M}(B)$, so that $\mathcal{M}(B)$ is homeomorphic to $sp_A(a)$. It follows that B may be identified with the algebra $C(sp_A(a))$ through which the element a corresponds to the function ι with $\iota(\lambda) = \lambda$ and a^* to $\bar{\iota}$ with $\bar{\iota}(\lambda) = \bar{\lambda}$.

Definition 2.2.2. Keep the notations above, and let f be a continuous function on the set $sp_A(a)$. We denote by $f(a)$ the corresponding element in B.

The above correspondence is usually called spectral calculus for a. We note that this definition is compatible with the definition of $f(a)$ in 2.1, when f satisfies the condition. For each element $h \in A_h$ we can then define those elements such as

$$|h| = (h^2)^{\frac{1}{2}} \quad , \quad h_+ = \frac{1}{2}(|h| + h) \quad \text{and} \quad h_- = \frac{1}{2}(|h| - h) \quad ,$$

which are called the absolute value of h, the positive and negative parts of h respectively. It follows that any self-adjoint element h of A is written as a difference of orthogonal "positive" elements h_+ and h_-. The next theorem characterizes those "positive" elements.

Theorem 2.1.3. For a self-adjoint element a of a C*-algebra A the following conditions are equivalent;

(1) $sp(a) \subset |0, \infty)$;
(2) $a = b^*b$ for some $b \in A$;
(3) $a = h^2$ for some $h \in A_h$.

The set A_+ of all such elements a's forms a closed convex cone of A such that $A_+ \cap (-A_+) = \{0\}$.

Proof. We shall only give the property of A_+ and the implication (2) $\Rightarrow$ (3) since others are easily derived from the discussion given before Definition 2.2.2. Let $A_+ = \{a \in A_h | sp(a) \subset [0, \infty)\}$. Let S be the unit ball of A. We then have

$$A_+ \cap S = \{a \in A_h \cap S | \| 1 - a \| \leq 1\}$$

as seen from (2.2.1). Since A_+ is invariant by multiples of positive scalars,

$$A_+ = \{a \in A_h | \| \|a\| - a \| \leq \|a\| \} \quad .$$

Hence A_+ is closed. Moreover, for any $a, b \in A_+ \cap S$

$$\| 1 - \frac{1}{2}(a + b) \| = \frac{1}{2} \| (1 - a) + (1 - b) \| \leq 1 \quad ,$$

whence $\frac{1}{2}(a + b) \in A_+ \cap S$. Thus $A_+ \cap S$ is convex, and A_+ is a convex cone. The property $A_+ \cap (-A_+) = \{0\}$ is easily seen from (2.3). Next suppose that $a = b^*b$ for some $b \in A$. We put $c = (a_+)^{\frac{1}{2}}$ and $d = (a_-)^{\frac{1}{2}}$. Then,

$$(bd)^*(bd) = d(b^*b)d = d(c^2 - d^2)d = -d^4 \in (-A_+)$$

by the implication (3) $\Rightarrow$ (1). Write $bd = h_1 + ih_2$ with $h_1, h_2 \in A_h$. We have

$$\begin{aligned}(bd)(bd)^* &= -(bd)^*(bd) + (h_1 - ih_2)(h_1 + ih_2) \\ &\quad + (h_1 + ih_2)(h_1 - ih_2) \\ &= -(bd)^*(bd) + 2h_1^2 + 2h_2^2 A_+\end{aligned}$$

because A_+ is shown to be a convex cone. Hence by (2.1.4)

$$(bd)^*(bd) \in A_+ \cap (-A_+) = \{0\} \quad .$$

Therefore $d^4 = 0$ and $d = 0$, whence $a = c^2$. This completes the proof.

With this convex cone A_+ we can consider an order in A_h defining $a \geq b$ for $a, b \in A_h$ by $a - b \in A_+$.

An approximate identity of a Banach algebra A is a net $\{u_\alpha\}$ of elements such that

$$\lim_\alpha \| u_\alpha a - a \| = \lim_\alpha \| a u_\alpha - a \| = 0 \quad , \quad a \in A \quad .$$

If $\{u_\alpha\}$ is bounded $\{u_\alpha\}$ is called a bounded approximate identity.

Theorem 2.2.4. Let A be a C*-algebra and S_0 be the open unit ball of A. If I is an ideal of A, then the set $S_0 \cap I_+$ is upward directed and forms an approximate identity for $\bar{I}$, the closure of I. In particular, even if a C*-algebra A does not have a unit it always has an approximate identity consisting of the elements in $S_0 \cap A_+$.

The following result shows an application of approximate identities.

Theorem 2.2.5. A closed ideal I of a C*-algebra A is self-adjoint and the quotient algebra A/I is also a C*-algebra.

Proof. Let $\{u_\alpha\}$ be the approximate identity of I. Then,

$$\lim_\alpha \| x^* - x^* u_\alpha \| = \lim_\alpha \| x - u_\alpha x \| = 0$$

for $x \in I$. As $x^* u_\alpha \in I$, x^* belongs to I, that is, I is self-adjoint. It follows that the quotient algebra A/I is a Banach *-algebra. In order to see the axiom (2.2) for A/I it is enough to show that

$$\| \hat{a}^* \hat{a} \| \geq \| \hat{a} \|^2 \qquad \hat{a} \in A/I \quad .$$

We assert first that $\| \hat{a} \| = \lim_\alpha \| a - a u_\alpha \|$. In fact, if $x \in I$

$$\begin{aligned} \limsup_\alpha \| a - a u_\alpha \| &= \limsup_\alpha \| (a + x)(1 - u_\alpha) \| \\ &\leq \| a + x \| \quad , \end{aligned}$$

which implies that

$$\|\hat{a}\| = \inf\{\|a+x\| \mid x\in I\}$$

$$\leq \liminf \|a - au_\alpha\| \leq \limsup \|a - au_\alpha\|$$

$$\leq \inf\{\|a+x\| \mid x\in I\} = \|\hat{a}\| \quad .$$

Therefore,

$$\|\hat{a}\|^2 = \lim\|a - au_\alpha\|^2 = \lim\|(a - au_\alpha)^*(a - au_\alpha)\|$$

$$= \lim\|(1 - u_\alpha)a^*a(1 - u_\alpha)\|$$

$$= \lim\|(1 - u_\alpha)(a^*a + x)(1 - u_\alpha)\|$$

$$\leq \|a^*a + x\| \quad .$$

Hence, $\|\hat{a}\|^2 \leq \|\hat{a}^*\hat{a}\|$.

As we shall see later in (2.3.1) any *-homomorphism π (i.e. $\pi(a^*) = \pi(a)^*$) of a C*-algebra into another C*-algebra is norm decreasing. Thus as an immediate consequence of the above theorem we obtain

Corollary 2.2.6. If π is a *-homomorphism of a C*-algebra A onto a C*-algebra B, then π induces canonically a *-isomorphism between the quotient C*-algebra $A/\pi^{-1}(0)$ and B.

We shall also see later in (2.3.1) that the image $\pi(A)$ of a C*-algebra A is always closed and the *-isomorphism between $A/\pi^{-1}(0)$ and $\pi(A)$ is necessarily isometric.

2.3 States and Representations of Banach *-algebras and C*-algebras

We begin with the following

Definition 2.3.1. A representation π of a Banach *-algebra A on a Hilbert space H is a *-homomorphism of A into the algebra $B(H)$. It is said to be non-degenerate if $|\pi(A)H| = H$ and cyclic if there exists a vector ξ_0 such that $|\pi(A)\xi_0| = H$. The vector ξ_0 is called a cyclic vector for π. We say a representation is faithful if it is injective.

For a representation π on H if a vector ξ is orthogonal to the subspace $|\pi(A)H|$, then $\pi(a)\xi = 0$ for every $a \in A$. Therefore, π acts essentially on the subspace $|\pi(A)H|$, so that we usually consider non-degenerate representations. Besides, it is also not so hard to see that any representation is a direct sum of cyclic representations where sum of representations will be understood in a obvious way.

The next result shows the automatic continuity of a representation of a Banach *-algebra.

Theorem 2.3.1. Let π be a *-homomorphism of a Banach *-algebra A into a C*-algebra B. Then we have

(1) $\| \pi(a) \| \leq \| a \|$;

(2) If A is a C*-algebra, the image $\pi(A)$ is closed in B and if moreover π is injective it is isometric.

Proof. (1) Since π extends to a homomorphism on A_1 if A is not unital, we may assume that A is unital. Let $B_0 = \pi(A)$, the closure of $\pi(A)$, then $sp_{B_0}(a) \subseteq sp_A(a)$. It follows by (2.3)

$$\| \pi(a) \|^2 = \| \pi(a)^*\pi(a) \| = \| \pi(a^*a) \|$$

$$= \| \pi(a^*a) \|_{sp} \leq \| a^*a \|_{sp} \leq \| a^*a \| \leq \| a \|^2 \quad ;$$

(2) Suppose first that π is injective. It suffices to show that $\| \pi(a^*a) \| = \| a^*a \|$. Therefore specializing the situation locally together with an adjustment as in the above proof of (1) we may confine the attention to the case where both A and B are commutative unital C*-algebras and $\pi(A) = B$. Let ${}^t\pi$ be the transpose map of B^* to A^*. Then ${}^t\pi(\Delta(B)) \subseteq \Delta(A)$ and if they are not equal there exists a nonzero continuous function a on $\Delta(A)$ (regarded as an element of A) vanishing on ${}^t\pi(\Delta(B))$. This however means that $\pi(a)$ vanishes on $\Delta(B)$ and $\pi(a) = 0$ by (2.2.1), a contradiction. Hence, ${}^t\pi(\Delta(B)) = \Delta(A)$

and again by (2.2.1) π is isometric with $\pi(A) = B$. In general, we note that the homomorphism π induces an isomomorphism $\hat{\pi}$ of the quotient C*-algebra $A/\pi^{-1}(0)$ into B. Therefore, the image $\hat{\pi}(A/\pi^{-1}(0)) = \pi(A)$ is closed in B. This completes the proof.

Definition 2.3.2. Let A be a Banach *-algebra. A linear functional φ on A is said to be positive if $\varphi(a^*a) \geq 0$ for every $a \in A$, selfadjoint if $\varphi(a^*) = \overline{\varphi(a)}$. We say that a positive linear functional ψ is majorized by φ if $\psi(a^*a) \leq \varphi(a^*a)$ for every $a \in A$.

Positive linear functionals play central rôle in the representation theory of Banach *-algebras, especially of C*-algebras. In this connection, particularity of positive cones of C*-algebras often reflects the properties of positive linear functionals. A positive linear functional becomes often automatically continuous as in the following theorem though the result is highly non-trivial.

Theorem 2.3.2. Let A be a Banach *-algebra with a bounded approximate identity, then every positive linear functional on A is continuous, i.e. bounded.

With this theorem we have

Proposition 2.3.3. Let φ be a positive linear functional on a Banach *-algebra A with a bounded approximate identity $\{u_\alpha\}$ of norm $\| u_\alpha \| \leq k$. Let φ be a positive linear functional on A, then

(1) φ is self-adjoint;

(2) $|\varphi(b^*a)|^2 \leq \varphi(b^*b)\,(a^*a)$ (Schwarz inequality), and
$|\varphi(a)|^2 \leq k^2 \|\varphi\| \varphi(a^*a)$.

Proof. (1) From the polarization identity

$$4b^*a = \sum_{n=0}^{3} i^n (a + i^n b)^*(a + i^n b) \qquad ,$$

we see that $\varphi(a^*b) = \overline{\varphi(b^*a)}$ in general. On the other hand, since φ is continuous by the above theorem

$$\varphi(a^*) = \lim \varphi(a^*u_\alpha) = \lim \overline{\varphi(u_\alpha^* a)}$$

$$= \lim \overline{\varphi((a^*u_\alpha)^*)} = \overline{\varphi(a)} \quad .$$

The first inequality in (2) follows from the fact

$$0 \le \varphi((\lambda a + \mu b)^*(\lambda a + \mu b))$$

$$= |\lambda|^2\varphi(a^*a) + 2\mathrm{Re}\bar{\lambda}\mu\varphi(a^*b) + |\mu|^2\varphi(b^*b)$$

for every scalars λ, μ. Hence,

$$|\varphi(a)|^2 = \lim|\varphi(u_\alpha a)|^2$$

$$\le \limsup \varphi(u_\alpha u_\alpha^*)\varphi(a^*a) \le k^2\|\varphi\|\varphi(a^*a) \quad .$$

<u>Proposition 2.3.4.</u> Let A be a unital Banach *-algebra and φ a linear functional.

(1) If φ is positive, then $\|\varphi\| = \varphi(1)$.

(2) If A is a C*-algebra and $\varphi(1) = \|\varphi\|$, then φ is positive

<u>Proof.</u> The assertion (1) is clear because by the above proposition we see that $\|\varphi\| \le \varphi(1)$.

(2) Take an element $a \in A^+$ and suppose that $\varphi(a) \notin |0, \infty)$. We may assume that $\varphi(1) = 1$. By (2.1.3) we can find a complex number λ_0 and a disk $D = \{\lambda \in \mathbb{C} \| \lambda - \lambda_0| \le r\}$ such that D contains $sp(a)$ but not $\varphi(a)$. We then see that $\|a - \lambda_0\| \le r$ by (2.3) but also

$$|\varphi(a) - \lambda_0| = |\varphi(a - \lambda_0)| \le \|\varphi\| \, \|a - \lambda_0\| \le r \quad ,$$

a contradiction. Thus, $\varphi(a) \ge 0$.

Before going to state positive extensions of positive linear functionals on a C*-algebra we need

<u>Lemma 2.3.5.</u> Let B be a C*-subalgebra of a C*-algebra A. Then a positive linear functional φ on B extends to a positive

linear functional $\tilde{\varphi}$ on a unital C*-subalgebra B_1 generated by B and the unit of A with the same norm by putting $\tilde{\varphi}(1) = \|\varphi\|$. In particular a positive linear functional on a nonunital C*-algebra extends to a positive linear functional on the C*-algebra obtained by adjoining a unit.

Proof. Let $\{u_\alpha\}$ be an approximate unit for B. We first assert that $\|\varphi\| = \lim \varphi(u_\alpha)$. Indeed, $\{\varphi(u_\alpha)\}$ is an increasing net in R_+ with limit $\lambda_0 \le \|\varphi\|$. On the other hand, by (2) of (2.3.3) for each element a in the unit sphere of B,

$$|\varphi(u_\alpha a)|^2 \le \varphi(u_\alpha)\varphi(a^*u_\alpha a) \le \lambda_0\|\varphi\| \quad ,$$

whence $\|\varphi\|^2 \le \lambda_0\|\varphi\|$ and $\|\varphi\| \le \lambda_0$. Therefore for each element $a \in B$ and a complex number λ,

$$\begin{aligned}\tilde{\varphi}((\lambda + a)^*(\lambda + a)) &= |\lambda|^2\tilde{\varphi}(1) + \bar{\lambda}\varphi(a) + \lambda\varphi(a^*) + \varphi(a^*a) \\ &= \lim_\alpha \varphi(|\lambda|^2u_\alpha + \bar{\lambda}u_\alpha a + \lambda a^*u_\alpha + a^*a) \\ &\ge \limsup_\alpha \varphi(|\lambda|^2u_\alpha^2 + \bar{\lambda}u_\alpha a + \lambda a^*u_\alpha + a^*a) \\ &= \limsup_\alpha \varphi((\lambda u_\alpha + a)^*(\lambda u_\alpha + a)) \ge 0 \quad ,\end{aligned}$$

that is, φ is positive and $\|\tilde{\varphi}\| = \tilde{\varphi}(1) = \|\varphi\|$.

Proposition 2.3.6. Let A be a C*-algebra (not necessarily unital) and B be a C*-subalgebra of A. Then a positive linear functional φ on B extends to a positive linear functional on A in a norm preserving way.

Proof. Let A_1 be the C*-algebra obtained by adjoining a unit to A if A is not unital. The subalgebra B is then regarded as a C*-subalgebra of A_1. Thus in any case we extend φ first to the unital C*-subalgebra B_1 of A_1 (or, A if A is unital) as a positive linear functional and then to the whole algebra by the Hahn-Banach extension theorem. By (2) of (2.3.4) this provides a required extension.

A positive linear functional φ on a Banach *-algebra A with a bounded approximate identity gives rise to a cyclic representation $\{\pi_\varphi, H_\varphi, \xi_\varphi\}$ called the GNS (Gelfand-Naimark-Segal)-representation. For this construction we need first the following

Lemma 2.3.7. The set

$$N_\varphi = \{x \in A | \varphi(x^*x) = 0\}$$

is a closed left ideal of A (called the left kernel of φ).

Proof. Let x, y be elements of N_φ and a be an element of A. We have then

$$\begin{aligned}\varphi((x+y)^*(x+y)) &= \varphi(x^*x + x^*y + xy^* + y^*y)\\ &= 2\mathrm{Re}\varphi(x^*y)\\ &\leq 2\varphi(x^*x)^{\frac{1}{2}}\,\varphi(y^*y)^{\frac{1}{2}} = 0 \quad ,\end{aligned}$$

and

$$\begin{aligned}\varphi((ax)^*(ax)) &= \varphi(x^*a^*ax)\\ &\leq \varphi(x^*x)^{\frac{1}{2}}\,\varphi((a^*ax)^*(a^*ax)) = 0 \quad .\end{aligned}$$

Hence both $x+y$ and ax lie in N_φ. The other properties of N_φ obviously hold.

Theorem 2.3.8. Let φ be a positive linear functional on a Banach *-algebra A with a bounded approximate identity. Then φ gives rise to, within unitary equivalence, a unique cyclic representation $\{\pi_\varphi, H_\varphi, \xi_\varphi\}$ of A with the following properties:

(1) $|\pi_\varphi(A)\xi_\varphi| = H_\varphi$,

(2) $\varphi(a) = (\pi_\varphi(a)\xi_\varphi|\xi_\varphi) \qquad a \in A$.

Proof. We equip first the quotient space A/N_φ by the inner product

$$(\eta_\varphi(x)|\eta_\varphi(y)) = \varphi(y^*x) \qquad x, y \in A \quad ,$$

where $\eta_\varphi(x)$ means the coset $x+N$. Define a linear operator $\pi_\varphi^0(a)$ for $a \in A$ on the pre-Hilbert space A/N_φ by

$$\pi_\varphi^0(a)\eta_\varphi(x) = \eta_\varphi(ax) \quad .$$

Then by (2.3.3)

$$\begin{aligned} |(\pi_\varphi^0(a)\eta_\varphi(x)|\eta_\varphi(y))| &= |\varphi(y^*ax)| \\ &\le \varphi(y^*y)^{\frac{1}{2}}\ \varphi(x^*a^*ax)^{\frac{1}{2}} \\ &\le \|\eta_\varphi(y)\|\ \|a\|\ \varphi(x^*x)^{\frac{1}{2}} \\ &= \|a\|\ \|\eta_\varphi(x)\|\ \|\eta_\varphi(y)\| \quad . \end{aligned}$$

The last inequality follows from the inequality, $x^*a^*ax \le \|a\|^2\ x^*x$. Therefore $\pi_\varphi^0(a)$ extends to a bounded operator $\pi_\varphi(a)$ on the Hilbert space H_φ, the completion of A/N_φ. It is then apparent that the map $a \in A \to \pi_\varphi(a)$ is a representation of A on H_φ.

Next consider the functional ω on A/N_φ defined by $\omega(\eta_\varphi(x)) = (x)$. With k for the bound of a bounded approximate identity for A we have by (2.3.3)

$$|\omega(\eta_\varphi(x))| \le k\|\varphi\|^{\frac{1}{2}}\ \|\eta_\varphi(x)\| \quad .$$

Hence ω extends to a bounded linear functional on H_φ and by the Riesz theorem there exists a unique vector $\xi_\varphi \in H_\varphi$ such that

$$\varphi(a) = \omega(\eta_\varphi(a)) = (\eta_\varphi(a)|\xi_\varphi) \quad .$$

On the other hand, for every $x \in A$

$$\begin{aligned} (\eta_\varphi(a)|\eta_\varphi(x)) &= \overline{\varphi(a^*x)} = \overline{(\eta_\varphi(a^*x)|\xi_\varphi)} \\ &= \overline{(\pi_\varphi(a)^*\ \eta_\varphi(x)|\xi_\varphi)} = (\pi_\varphi(a)\xi_\varphi|\eta_\varphi(x)) \quad . \end{aligned}$$

Hence,

$$\eta_\varphi(a) = \pi_\varphi(a)\xi_\varphi \quad .$$

Finally, let $\{\pi'_\varphi, H'_\varphi, \xi'_\varphi\}$ be another representation of A with the same properties. Define a map u_0 of the subspace $\pi'_\varphi(A)\xi'_\varphi$ to the subspace $\pi_\varphi(A)\xi_\varphi$ by

$$u_0\pi'_\varphi(x)\xi'_\varphi = \pi_\varphi(x)\xi_\varphi \qquad x \in A \quad .$$

Since

$$\begin{aligned}(u_0\pi'_\varphi(x)\xi'_\varphi | u_0\pi'_\varphi(y)\xi'_\varphi) &= (\pi_\varphi(x)\xi_\varphi | \pi_\varphi(y)\xi_\varphi) \\ &= (\pi_\varphi(y^*x)\xi_\varphi | \xi_\varphi) \\ &= \varphi(y^*x) = (\pi'_\varphi(y^*x)\xi'_\varphi | \xi'_\varphi) \\ &= (\pi'_\varphi(x)\xi'_\varphi | \pi'_\varphi(y)\xi'_\varphi) \quad ,\end{aligned}$$

the map u_0 is well defined and an isometry. Therefore we can extend u_0 to a unitary map u of $H_{\varphi'}$ to H_φ. One may then easily check that u impliments the unitary equivalence of the representations π'_φ and π_φ. This completes the proof.

We note that every cyclic representation arises in this way.

<u>Definition 2.3.3.</u> A representation π of a Banach *-algebra A on a Hilbert space H is said to be irreducible if there is no proper $\pi(A)$-invariant closed subspace of H.

As it is easily verified a projection to a $\pi(A)$-invariant closed subspace commutes with every element of $\pi(A)$, hence we can say that π is irreducible if and only if there is no projections in H commuting with $\pi(A)$ except 0 or 1. Combining this result with resolutions of self-adjoint operators one can conclude that π is irreducible if and only if there is no operators commuting with $\pi(A)$ except scalar operators.

<u>Definition 2.3.4.</u> Let A be a Banach *-algebra with a bounded

approximate identity. We call a positive linear functional φ on A a state if $\|\varphi\| = 1$. The set of all states on A, S(A), is called the state space of A. The functional φ is also said to be pure if every positive linear functional ψ such that $0 \leq \psi \leq \varphi$ is of the form $\psi = \lambda\varphi$ for $0 \leq \lambda \leq 1$. We denote P(A) the set of all pure states on A.

Let Q_A be the set of all positive linear functionals with norm not greater than 1. Then Q_A is apparently a $\sigma(A^*, A)$-compact convex subset of the unit sphere of A^*. One easily verifies that a pure state is nothing but an extreme point of Q_A. When A is unital, the state space S(A) becomes also a $\sigma(A^*, A)$-compact convex set. One can show that a state is pure if and only if it is an extreme point of S(A). In this case the Krein-Milman theorem says that S(A) is the closed convex hull of P(A).

<u>Proposition 2.3.9.</u> Let φ be a positive linear functional on a Banach *-algebra A with a bounded approximate identity. Then the GNS-representation $\{\pi_\varphi, H_\varphi, \xi_\varphi\}$ is irreducible if and only if φ is pure.

<u>Proof.</u> Let φ be pure and suppose that K were a proper invariant subspace of H_φ with the projection p on K. Define a positive linear functional

$$\psi(a) = (\pi_\varphi(a)p\xi_\varphi | \xi_\varphi) .$$

Since p commutes with every operator $\pi_\varphi(a)$ we have

$$\psi(a^*a) = \| \pi_\varphi(a)p\xi_\varphi\|^2 = \| p\pi_\varphi(a)\xi_\varphi\|^2$$

$$\leq \| \pi_\varphi(a)\xi_\varphi\|^2 = \varphi(a^*a) \qquad a \in A .$$

Hence $\psi = \lambda\varphi$ with $0 \leq \lambda \leq 1$. One then sees that for $x, y \in A$

$$(\lambda\pi_\varphi(x)\xi_\varphi | \pi_\varphi(y)\xi_\varphi) = (p\pi_\varphi(x)\xi_\varphi | \pi_\varphi(y)\xi_\varphi) ,$$

whence $p = \lambda 1$, a contradiction. Conversely take a positive linear functional ψ with $\psi \leq \varphi$. We define a semi-inner product on the subspace $\{\pi_\varphi(x)\xi_\varphi | x \in A\}$ by

$$< \pi_\varphi(x)\xi_\varphi | \pi_\varphi(y)\xi_\varphi > = \psi(y^*x) \quad .$$

As $0 \leq \psi \leq \varphi$, it is well defined and extends to a semi-inner product on the whole space H_φ with

$$< \xi | \xi > \leq \| \xi \|^2 \quad .$$

Hence there exists an operator a such that $0 \leq a \leq 1$ and

$$< \xi | \eta > = (a\xi | \eta) \quad .$$

By a straightforward calculation we obtain,

$$(a\pi_\varphi(x)\pi_\varphi(y)\xi_\varphi | \pi_\varphi(z)\xi_\varphi) = (\pi_\varphi(x)a\pi_\varphi(y)\xi_\varphi | \pi_\varphi(z)\xi_\varphi)$$

for every x, y and z in A. Thus the operator a commutes with $\pi_\varphi(A)$ and $a = \lambda 1$. Therefore,

$$\begin{aligned} \psi(x) &= \lim_\alpha \psi(u_\alpha x) \\ &= \lim_\alpha < \pi_\varphi(x)\xi_\varphi | \pi_\varphi(u_\alpha^*)\xi_\varphi > \\ &= \lim_\alpha (a\pi_\varphi(x)\xi_\varphi | \pi_\varphi(u_\alpha^*)\xi_\varphi) \\ &= \lim_\alpha (a\pi_\varphi(u_\alpha x)\xi_\varphi | \xi_\varphi) = \lambda \lim_\alpha (\pi_\varphi(u_\alpha x)\xi_\varphi | \xi_\varphi) \\ &= \lambda\varphi(x) \quad . \end{aligned}$$

It is to be noticed that since an irreducible representation of $C(X)$ is one dimensional a pure state on $C(X)$ must be multiplicative and it is a character. Conversely, a character on $C(X)$ is clearly a pure state. With this identification the Stone-Weierstrass approximation theorem reads that a unital C*-subalgebra of $C(X)$ separating the pure states of $C(X)$ coincides with the algebra $C(X)$. The non-commutative Stone-Weierstrass version of this theorem is known as the Stone-Weierstrass theorem for C*-algebras: namely, if B is a unital C*-subalgebra of a C*-algebra A separating the pure states of A, then $A = B$. Though there are many partial results about the theorem the complete answer is one of the outstanding problem in the theory of C*-algebras. Thus we do not know whether the theorem is true or not in general.

We know by (2.3.6) that a state φ on a C*-subalgebra B of a C*-algebra A extends to a state on A. Moreover, if φ is a pure state it is also extended to a pure state on A. In fact, the extension $\tilde{\varphi}$ in (2.3.5) is pure if φ is pure and a pure state extension of $\tilde{\varphi}$ to A is obtained as an extreme point of the set of all state extensions of $\tilde{\varphi}$ to A.

We shall show next the main representation.

<u>Theorem 2.3.10.</u> A C*-algebra A admits a faithful representation and it also admits sufficiently many irreducible representations.

The theorem establishes the equivalence between the concept of abstract C*-algebras and that of concrete C*-algebras. Henceforth we may always assume, if needed, that A is acting on a suitable Hilbert space.

<u>Proof.</u> Take a non-zero element a, then $-a^*a \notin A_+$. Since A_+ is a closed convex cone in the real Banach space A_h, there exists by the Hahn-Banach theorem a real linear functional φ_a on A_h separating A_+ and $-a^*a$. Hence, $\varphi_a(-a^*a) < 0$. We extend φ_a to A by defining

$$\hat{\varphi}(x + iy) = \varphi_a(x) + i\varphi_a(y) \qquad x, y \in A_h \quad .$$

The extended functional $\hat{\varphi}$ is apparently positive and $\hat{\varphi}(a^*a) > 0$. Let $\{\pi_a, H_a, \xi_a\}$ be the GNS-representation of A by $\hat{\varphi}$. Then we have $\pi_a(a) \neq 0$ because

$$\begin{aligned} \| \pi_a(a)\xi_a \|^2 &= (\pi_a(a^*a)\xi_a | \xi_a) \\ &= {}_a(a^*a) > 0 \quad . \end{aligned}$$

Considering the direct sum representation π of these representations π_a's over all non-zero elements of A we see that A admits the faithful representation π.

The above argument exactly shows that there are sufficiently many states on A_+. Thus for the latter assertion it suffices to see that there are sufficiently many pure states on A_+, but this follows simply from the Krein-Milman theorem as we have mentioned before. This completes the proof.

Let A be a Banach *-algebra and let $\{\pi_\alpha\}$ be the set of all representations of A. Since every representation of A is norm decreasing we have

$$\| a \|_\infty = \sup \| \pi_\alpha(a) \| \leq \| a \| \qquad a \in A \quad .$$

Now suppose that A has sufficiently many representations (namely for any non-zero a there exists a representation π with $\pi(a) \neq 0$, then $\| a \|_\infty$ becomes a C*-norm on A.

<u>Definition 2.3.5.</u> We call the completion of A with respect to the norm $\| a \|_\infty$ the C*-envelope of A and denote by C*(A).

The convolution algebra $L^1(G)$ for a locally compact group G is an example of a Banach *-algebra with the above property (cf. 3.1). In this case, a faithful representation π on the Hilbert space $L^2(G)$ is obtained as

$$(\pi(f)g)(t) = \int_G f(s)g(s^{-1}t)ds \quad .$$

In general, if there exist sufficiently many positive linear functionals on the positive portion of a Banach *-algebra as in the case of C*-algebras, the algebra satisfies the condition. We shall be deeply concerned with one of such algebra, $\ell^1(G, A)$ for a discrete group G and a C*-algebra A. This is the algebra with convolution product twisted by an action of G on A as a group of *-automorphisms of A.

APPENDIX B

VON NEUMANN ALGEBRAS

B.1. Factors and their Types

Let H be a Hilbert space. In the algebra B(H) there are various kinds of locally convex topologies besides the operator norm topology. Among them the weak *-topology with respect to the duality $T(H)^* = B(H)$ where T(H) is the space of all trace class operators with trace class norm is notable because the unit sphere of B(H) becomes compact in that topology. This topology is usually called the σ-weak topology and coincides with the weak operator topology on the unit sphere of B(H). A von Neumann algebra M is then defined as a σ-weakly closed C*-subalgebra of B(H) containing the identity operator. But it turns out that in order to define a von Neumann algebra one may use any of other topologies except the norm topology. Thus, M is a particular C*-algebra which is by definition the conjugate space of the Banach space $M_* = M^0 \cap T(H)$. This special feature actually characterizes such a C*-algebra. Namely, a C*-algebra which is the conjugate space of a Banach space can be faithfully represented as a von Neumann algebra on some Hilbert space. The concept of von Neumann algebras sticks nevertheless heavily on their acting spaces as shown in the following fundamental theorem of von Neumann.

Let M' be the set of all operators in B(H) commuting with every element of M. The set M' is called the commutant of M and it becomes naturally a von Neumann algebra. Apparently we have $M \subset M''$

and $M' \subset M'''$.

Theorem B.1. The following assertions are equivalent;

(1) M is a von Neumann algebra on a Hilbert space H;

(2) $M = M''$.

Thus discussions of M are necessarily concerned with the pair (M, M'), so that we refer a von Neumann algebra as $\{M, H\}$.

Now as a consequence of the theorem one can see that a von Neumann algebra contains all spectral projections of every self-adjoint element. Hence, M is rich in projections contrary to usual C*-algebras. Every non-degenerate representation of a C*-algebra (regardless of unital or nonunital case) gives rise to a von Neumann algebra if we close up the represented C*-subalgebra of $B(H)$ in the σ-weak topology. One could say that the theory of von Neumann algebras is the representation theory of C*-algebras. Thus, in particular, abundance of instances of von Neumann algebras should be cited but the trouble is at the point that we need somewhat heavy preparation to recognize them precisely as specified examples for various types of von Neumann algebras. A von Neumann algebra with trivial center is called a factor and a representation of a C*-algebra which generates a factor is called a factor representation.

On the other hand, a commutative von Neumann algebra is regarded as the representation of the algebra $L^{\infty}(X, \mu)$ for a suitable measure space (X, μ) represented on the space $L^{2}(X, \mu)$ as multiplication operators.

There is a theory of integration of the fields of operators having their values in the fields of factors over separable Hilbert spaces. When the acting space H is separable, a von Neumann algebra M is decomposed into an integration over such a field of factors together with the decomposition of H. Thus, in principle, we may confine our study to factors. Henceforth, M stands mostly a factor on a separable Hilbert space H. A function τ defined on the positive cone M_+ with extended value $[0, \infty]$ is called a trace of M if it satisfies the following conditions:

(1) $\tau(\lambda a + \mu b) = \lambda\tau(a) + \mu\tau(b) \qquad \lambda, \mu \geq 0$,

(2) $\tau(a^*a) = \tau(aa^*)$.

A trace τ is said to be normal if $\tau(\sup a_\alpha) = \sup \tau(a_\alpha)$ for every directed increasing family of positive elements of M and faithful if $\tau(a) = 0$ implies $a = 0$. It is said to be finite if $\tau(1)$ is finite and semi-finite if for every nonzero element $a \in M_+$ there exists a nonzero element b such that $0 \leq b \leq a$ and $\tau(b)$ is finite. If a trace τ is finite it extends to the whole algebra M by linearity. This extended positive linear functional becomes necessarily normal and faithful on the factor M and satisfies the relation $\tau(xy) = \tau(yx)$ for $x, y \in M$. Furthermore in this case, τ is unique up to scalar constants. Thus, a faithful normal semi-finite trace is also unique up to constants. The factor M is then classified according to the existence of various kinds of faithful normal traces on M with the following ranges on the set of projections;

(1) $\{1,2,\ldots,n\}$ or $\{1,2,\ldots,\infty\}$;

(2) The interval $[0, 1]$ (τ is normalized as $\tau(1) = 1$) or $[0,\infty]$.

(3) No finite value except zero.

M is then said to be of type I_n and I_∞ respectively in case (1) of type II_1 (finite type II) and II_∞ respectively in case (2) and of type III in case (3). Both cases (1) and (2) mean the existence of the finite or semi-finite trace. A factor of type I or type II is also said to be semi-finite. A factor of type I is characterized as the factor which is *-isomorphic to the algebra B(H).

There is another way to understand the types of von Neumann algebras. Two projections p and q in M are said to be equivalent, written as $p \sim q$, if there exists a partial isometry v in M such that $v^*v = p$ and $vv^* = q$. If there exists a subprojection q' of q which is equivalent to p we write that $p \lesssim q$. Then the comparability theorem naturally holds: namely, if $p \lesssim q$ and $q \lesssim p$, then $p \sim q$. A projection p is said to be finite if no subprojection of p in M is equivalent to p. The factor M is said to be finite if the identity projection is finite. It is then known that a faithful normal trace determines equivalences of projections. In particular, the finiteness is equivalent to the existence of the finite trace. In case of a semi-finite factor every nonzero projection contains a nonzero finite subprojection, whereas no nonzero finite projection exists in a factor of type III.

For a long time very few factors were recognized as factors of type II and type III, but as of now continuously many factors are specified as examples of non-*-isomorphic factors of type II and III. For such discussions developments of the theory from the side of mathematical physics such as quantum statistical mechanics cannot be overestimated.

One of the common way to provide examples of factors is to use the context of crossed products of the L^∞-algebras with respect to the ergodic group actions of measurable transformations. Let (X, μ) be the Lebesque measure space and G be a discrete group of non-singular measurable transformations on X. Then G induces an action α on $L^\infty(X, \mu)$. In this case, an automorphism α_s is said to be free if any nonzero projection p contains a nonzero subprojection q with $q\alpha_s(q) = 0$. If α_s is free for every α_s $(s \neq e)$ we call the action α free. Assume now that G is freely acting on the algebra $L^\infty(X, \mu)$, then the von Neumann crossed product $L^\infty(X, \mu) \rtimes_\alpha G$, defined as the weak closure of the reduced C*-crossed product $L^\infty(X, \mu) \rtimes_{\alpha r} G$ (cf. Chap. 3) becomes a factor if and only if α is ergodic.

Proposition B.2. Keep the above assumptions and set $M = L^\infty(X, \mu) \rtimes_\alpha G$.

(1) M is of type II_1 (resp. II_∞) if and only if there exists a G-invariant finite (resp. σ-finite but not finite) measure equivalent to μ;

(2) M is of type III if and only if there exists no G-invariant σ-finite measure equivalent to μ.

Examples of transformation groups of these cases are; for the case of type II_1 factors, the group of irrational rotation $\{n\theta\}$ and that of all rational numbers acting on the torus T with the Lebesque measure. For factors of type II_∞ it is enough to shift above situation to the real line R. To construct a factor of type III, the transformation group G on R consisting of those maps $\{\sigma_{a,b} | a > 0,\ a,\ b \text{ rational}\}$ defined as $\sigma_{a,b}(x) = ax + b$, is the simplest example. On the other hand, as the simplest case of crossed products the group von Neumann algebras, the weak closures of reduced group C*-algebras, offer examples of von Neumann algebras. For instance, a countable ICC group G provides a factor of type II_1 where G is called an ICC group if the conjugate class of every element $s\ (\neq e)$ is infinite. The permutation group of the natural integers N leaving but a finite number of integers fixed gives rise to the distinguished II_1-factor R_0 called the hyperfinite factor. A hyperfinite factor is originally defined as the weak closure of an increasing sequence of matrix algebras. Namely, it is the weak closure of a (concrete) UHF algebra (cf. Appendix C) but the situation here is much different from UHF algebras. The II_1-factor R_0 is known to be unique up to *-isomorphism. The free group on two generators provides another example of a II_1-factor.

In connection with representations of C*-algebras, the infinite C*-tensor product of 2×2-matrix algebras $A = \bigotimes_{n=1}^{\infty} M_2$ plays a very important rôle. Let Tr be the trace on M_2. For $0 \leq \lambda \leq 1$, consider a state $\varphi_\lambda = \bigotimes_{n=1}^{\infty} \varphi_\lambda^0$ of product type where φ_λ^0 is the state of M_2 defined as $\varphi_\lambda^0(a) = \mathrm{Tr}(ta)$ for the matrix $t = \begin{pmatrix} \frac{\lambda}{1+\lambda} & 0 \\ 0 & \frac{1}{1+\lambda} \end{pmatrix}$.

The GNS-representation of φ_λ yields then a hyperfinite factor M (called Powers factor). When $\lambda = 1$, it coincides with the factor R_0 whereas the state φ_0 gives rise to a factor of type I_∞. However, for $0 < \lambda < 1$ the factors M_λ turn out to be factors of type III and they are mutually not *-isomorphic.

B.2. Tomita-Takesaki Theory and Recent Developments

So far the first stage of the theory of von Neumann algebras was developed centering about the concept of the trace such as the equivalence of the finiteness by the existence of the finite trace and by the property of the identity projection. One of big use of the trace is then the standard representation of a semi-finite factor. Namely, a semi-finite factor $\{M, H\}$ can be faithfully represented on a special Hilbert space H_0 (simply the usual GNS-representation of the trace if it is finite) in which there exists an involution J (conjugate linear isometry with $J^2 = 1$) satisfying $JMJ = M'$. Within this strategy, however, there is no way to attack the structure of type III factors. In seeking new ways, one is guided to study group von Neumann algebras of non-unimodular groups and crucial breakthrough was made and amplified by M. Tomita and M. Takesaki (Tomita-Takesaki theory). In the simplest case where a factor $\{M, H\}$ has a separating and generating unit vector ξ_0 (separating if $a\xi_0 = 0$ implies $a = 0$ for $a \in M$ and generating if $|M\xi_0| = H$), the theory consists of the following discussions. We notice first that each faithful normal state φ on M realizes this situation through its (isomorphic) GNS-representation. In the space H, the densely defined conjugate linear operator: $a\xi_0 \to a^*\xi_0$ is closable and the polar decomposition of the closure $S = J\Delta^{\frac{1}{2}}$ gives rise to an involution J and a self-adjoint operator Δ.

<u>Theorem B.3.</u> (Tomita) Keep the notations as above. Then,

(1) $JMJ = M'$ (M is in the standard form on H);
(2) The one-parameter family of unitaries $\{\Delta^{it} | t \in R\}$ defines a strongly continuous one-parameter group of auto-morphisms of M, $\sigma_t = \mathrm{ad}\Delta^{it}$.

This group is called the modular automorphism group. Takesaki then proved that the automorphism group $\{\sigma_t\}$ is characterized by means of the KMS condition (Kubo-Martin-Schwinger, names after the study of quantum statistical mechanics) with respect to the state $\varphi_0(a) =$

$(a\xi_0, \xi_0)$. A one-parameter group $\{\alpha_t\}$ of automorphisms of M satisfies the KMS condition with respect to a faithful normal state φ if for every $a, b \in M$ there exists a complex valued function F which is bounded and continuous on the strip

$$\{z \in C \mid 0 \leq \operatorname{Im} z \leq 1\}$$

and analytic in its interior, such that

$$F(t) = \varphi(\alpha_t(a)b) \qquad F(t+i) = \varphi(b\alpha_t(a)) \qquad .$$

It was also shown that M is semi-finite if and only if every modular automorphism group is inner.

A weight on a factor $\{M, H\}$ is a function ϕ: $M_+ \to [0, \infty]$ satisfying

$$\phi(\lambda a + \mu b) = \lambda\phi(a) + \mu\phi(b) \qquad \lambda, \mu \geq 0 \qquad .$$

One can then define the extended GNS-representation π_ϕ (necessarily *-isomorphism) for a faithful normal semi-finite weight and obtain the same results as in the case of faithful normal states. The rôle of the semi-finite trace on a factor is now clarified as the weight having trivial modular automorphism group.

The concept of crossed product of von Neumann algebras offers not only a good way to construct examples of factors but also a basic instrument for the structure theory as typically shown in the following Takesaki's duality theorem.

Theorem B.4. Let α be an action of a locally compact abelian group G on a von Neumann algebra $\{M, H\}$. There is then an action $\hat{\alpha}$ (called the dual action of α) of the dual group $\hat{G}$ on the crossed product $M \rtimes_\alpha G$ such that the double crossed product $(M \rtimes_\alpha G) \rtimes_{\hat{\alpha}} \hat{G}$ is *-isomorphic to the tensor product $M \otimes B(L^2(G))$. In particular, if M is an infinite factor (and G is separable), M is *-isomorphic to $(M \rtimes_\alpha G) \rtimes_{\hat{\alpha}} G$.

The structure of type III factors is determined through crossed products by A. Connes and M. Takesaki. Every factor of type III, M is expressed as $N \rtimes_{\theta} R$ for a (uniquely determined) II_{∞} von Neumann algebra N (not necessarily a factor) and a faithful normal semi-finite trace τ for which $\tau \cdot \theta_t = e^{-t}\tau (t \in R)$. As of now, factors of type III are parametrized by numbers $0 \leq \lambda \leq 1$ and for a factor of type III_{λ} $(0 < \lambda < 1)$ another discrete decomposition by crossed product using single automorphism is known. Power factors $\{M_{\lambda} | 0 < \lambda < 1\}$ cited before are the family of hyperfinite factors of type III_{λ}.

It was long standing problems whether or not hyperfinite II_{∞}-factors are unique (hence has the form $R_0 \otimes B(H)$) and whether every infinite subfactor of R_0 is also hyperfinite (hence *-isomorphic to R_0). This was solved by Connes in the paper (Ann. Math. 104 (1976)) through the characterization of injective factors. A factor {M, H} is said to be injective if there exists a projection of norm one from B(H) to M.

Theorem B.5.

(1) A factor M on a separable Hilbert space H is injective if and only if it is hyperfinite;

(2) The hyperfinite factors of type II_{∞} and type III_{λ} $(0 < \lambda < 1)$ are unique (hence only Powers factors for type III_{λ} cases).

Since any subfactor of R_0 is shown to be injective the second problem mentioned above is an immediate consequence of the theorem.

There are many hyperfinite factors of type III_0, whereas the uniqueness of hyperfinite factors of type III_1 has been established by U. Haagerup.

CHAPTER 3

TRANSFORMATION GROUP C*-ALGEBRAS: CROSSED PRODUCTS OF COMMUTATIVE C*-ALGEBRAS

The second half of this chapter is the beginning of the interplay between topological dynamical systems and their associated C*-algebras. We introduce the C*-algebra A_Σ associated with a topological dynamical system $\Sigma = (X, G, \sigma_s)$ for a discrete group G, which we call the transformation group C*-algebra. We then discuss pure state extensions of the pure states on C(X) to A_Σ and also about tracial state extensions of invariant measures on X.

3.1. Group C*-algebras and Positive Definite Functions

In this section we briefly sketch the results about group C*-algebras. The concepts and results appearing here are prototypes of those in the theory of crossed products of C*-algebras when the space X is reduced to one point.

Let G be a locally compact unimodular group with the (left) Haar measure ds on G. Let $L^1(G)$ be the set of all integrable functions on G with the convolution and the *-operation:

$$f \times g(t) = \int f(s)g(s^{-1}t)ds = \int f(ts)g(s^{-1})ds \quad ,$$

$$f^*(t) = \overline{f(t^{-1})} \quad .$$

We denote M(G) the Banach space of all bounded complex Radon measures on G, identified with $C_0(G)^*$. The space $M(G)$ becomes a Banach *-algebra with respect to the following operations;

$$\int f(s)d(\mu\times\nu)(s) = \iint f(ts)d\mu(t)d\nu(s) \quad ,$$

$$\int f(s)d\mu^*(s) = \int f^*(s)d\mu(s) \quad .$$

We may then identify $L^1(G)$ as a closed self-adjoint ideal of $M(G)$ consisting of those measures that are absolutely continuous with respect to the Haar measure ds. For each s in G let δ_s be the point measure at s. We then have

$$\delta_s \times f(t) = f(s^{-1}t) \quad , \quad (f\times\delta_s)(t) = f(ts^{-1}) \quad .$$

The measure δ_e for the unit e is the unit of $M(G)$, whereas $L^1(G)$ has a unit only if G is discrete, though it has a bounded approximate identity. We call a unitary representation $\{u, H\}$ of G a homomorphism: $s\to u_s$ of G into the unitary group $U(H)$ of $B(H)$ which is continuous in the strong operator topology in $B(H)$.

<u>Proposition 3.1.1.</u> There are bijective correspondence among the sets of unitary representations of G, representations of $M(G)$ whose restrictions to $L^1(G)$ are non-degenerate, and non-degenerate representations of $L^1(G)$.

The correspondences are given by

$$(\pi(\mu)\xi|\eta) = \int (u_s\xi|\eta)d\mu(s) \qquad \mu\in M(G) \quad ,$$

$$(\pi(f)\xi|\eta) = \int f(s)(u_s\xi|\eta)ds \qquad f\in L^1(G) \quad .$$

We call the representation λ of G on $L^2(G)$ defined as

$$\lambda_s f(t) = f(s^{-1}t)$$

the left regular representation of G. The corresponding representation λ of $M(G)$ on $L^2(G)$ is defined as

$$\lambda(\mu)f(t) = \int f(s^{-1}t)d\mu(s) \quad .$$

It is faithful as well as its restriction to $L^1(G)$.

<u>Definition 3.1.1.</u> The C*-envelope of $L^1(G)$ is called the group C*-algebra of G, denoted by $C^*(G)$. The reduced C*-algebra $C_r^*(G)$ is the C*-algebra generated by the left regular representation λ of G.

We must remark that in spite of the convention of our discussions in this book these group C*-algebras are not unital unless G is discrete. As a concrete C*-algebra we can say that $C^*(G)$ is the norm closure of $\pi_u(L^1(G))$ where π_u is the universal representation (direct sum of all non-degenerate representations of $L^1(G)$), whereas $C_r^*(G)$ is the norm closure of $\lambda(L^1(G))$ on $L^2(G)$. Thus for a representation of G there always exists the corresponding representation of $C^*(G)$ on the same space.

If G is abelian, $C^*(G) = C_0(\hat{G})$ where $\hat{G}$ is the dual group of G. This will be seen from the fact that $C^*(G)$ is a commutative C*-algebra and the irreducible representations of $L^1(G)$ (hence those of $C^*(G)$) are one-dimensional, hence they correspond to the points of $\hat{G}$, the characters. Now consider $C^*(G)$ as the C*-algebra on the universal representation space H_u, then by (3.1.1) the map $s \to \delta_s$ is considered as a unitary representation of G with the property

$$\delta_s C^*(G) + C^*(G)\delta_s \subset C^*(G) \qquad s \in G \quad .$$

Let φ be a bounded linear functional on $C^*(G)$. Then although δ_s does not belong to $C^*(G)$ in general, by the above property we can extend φ in a unique way to a certain C*-algebra containing $C^*(G)$ and $\{\delta_s | s \in G\}$, the so-called multiplier algebra $M(C^*(G))$, so that we can define a function $\Phi(s) = \varphi(\delta_s)$. The function Φ turns out to be continuous on G and bounded by $\|\varphi\|$. Besides we have for

$f \in L^1(G)$

$$\varphi(f) = \int \varphi(\delta_s)f(s)ds = \int \Phi(s)f(s)ds \quad .$$

Therefore the map $\varphi \to \Phi$ is injective. When φ is positive, Φ is called a positive definite function because of the following properties.

Proposition 3.1.2. For each bounded continuous function Φ on G the following conditions are equivalent;

(1) Φ is positive definite;

(2) $\int \Phi(s)d(\mu^* \times \mu)(s) \geq 0$ for every $\mu \in M(G)$;

(3) For each finite set $\{s_i\}$ in G the matrix $|\Phi(s_i^{-1}s_j)|$ is positive definite.

For the proof we note first that the implication (1)$\Rightarrow$(2) is immediate as Φ defines a positive linear functional on $M(G)$. The assertion (2)$\Rightarrow$(3) follows by putting $\mu = \Sigma\lambda_i\delta_{s_i}$ for complex numbers $\{\lambda_i\}$. The assertion (3)$\Rightarrow$(1) is obtained through the procedure showing first that Φ defines a continuous positive linear functional φ on the algebra $K(G)$, the algebra of continuous functions with compact supports. We then extend φ to $M(G)$, thus concluding positive definiteness of Φ.

If G is abelian a positive definite function is precisely the Fourier transform of a bounded positive measure on $\hat{G}$, i.e. $\Phi(s) = \int_{\hat{G}} \langle s, \gamma \rangle d\mu(\gamma)$.

We shall next discuss when the regular representation of $C^*(G)$ is faithful. We need first the notion of amenability. For a function f on G denote by $\lambda_s f$ the function $t \to f(s^{-1}t)$. We say that f is left uniformly continuous if $\| \lambda_s f - f\|_\infty \to 0$ as $s \to e$. Let $C^b(G)$ the algebra of bounded continuous functions on G and $UC_l^b(G)$ the algebra of left uniformly continuous functions in $C^b(G)$. We define the algebra $UC_r^b(G)$ as the set of functions f's in $C^b(G)$ such that $f^* \in UC_\ell^b(G)$. Write

$$UC^b(G) = UC^b_\ell(G) \cap UC^b_r(G) \quad .$$

Let A be a C*-subalgebra of $L(G)$. We call a state m on A a left invariant mean if $m(\lambda_s f) = m(f)$ for every f in A. We then have

Proposition 3.1.3. A locally compact (unimodular) group G is said to be amenable if one of the following equivalent conditions hold;

(1) There is a left invariant mean on $UC^b(G)$;
(2) There is a left invariant mean on $C^b(G)$;
(3) There is a left invariant mean on $L^\infty(G)$.

Every abelian group is amenable as well as every compact group. Every closed subgroup of an amenable group is amenable. The free group on two generators is not amenable and it is a long remained problem whether a discrete group fails to be amenable only if it contains such a group as a subgroup.

Amenability of groups concerns with the present context at the point that if G is amenable there exists a net $\{\Phi_\alpha\}$ of certain positive definite functions in $K(G)$ with $\Phi_\alpha(e) = 1$ such that $\{\Phi_\alpha\}$ converges to 1 uniformly on compact subsets of G. Actually with this property one can prove the following

Theorem 3.1.4. The regular representation is faithful on $C^*(G)$ if and only if G is amenable.

Thus in this case the algebra $C^*(G)$ coincides with $C^*_r(G)$.

3.2 Crossed Products of C*-algebras

Let G be a discrete group and A a C*-algebra (assuming always a unit). Let α be an action of G on A, that is, α is a homomorphism of G into the group of all *-automorphisms of A, Aut A. We recall first that G is unimodular, though not amenable in general. A system $\{A, G, \alpha\}$ is called a C*-dynamical system. Let $k(G, A)$ be the algebra of all A-valued functions with compact (i.e. finite) support endowed with the following twisted convolution as product, involution and norm:

$$x^*(s) = \alpha_s(x(s^{-1})^*) \quad ,$$

$$xy(t) = \sum_s x(s)\alpha_s(y(s^{-1}t)) \quad ,$$

$$\| x \|_1 = \sum_s \| x(s) \| \quad .$$

The algebra $k(G, A)$ then becomes a normed algebra and we denote its completion by $\ell^1(G, A)$, a unital Banach *-algebra. The algebra A is regarded as a subalgebra of $k(G, A)$ with the same unit regarding each element a as a function on G with $a(e) = a$ and $a(s) = 0$ for $s \neq e$. Let E be the projection of $\ell^1(G, A)$ to this embedded subalgebra A such that $E(x)(e) = x(e)$. The map E is apparently continuous in norm and satisfies the identity:

$$E(axb) = aE(x)b \quad \text{for} \quad a, b \in A \quad .$$

Moreover,

$$\begin{aligned} E(x^*x)(e) &= x^*x(e) \\ &= \sum_s x^*(s)\alpha_s(x(s^{-1})) \\ &= \sum_s \alpha_s(x(s^{-1})^*x(s^{-1})) \quad , \end{aligned}$$

where the convergence is assured in norm. Hence E is a faithful positive projection in the sense that $E(x^*x) = 0$ implies $x = 0$. It follows from (2.3.5) that the algebra $\ell^1(G, A)$ has sufficiently many states on the positive portion. Therefore it has a faithful representation as mentioned at the end of Chap. 2.

Definition 3.2.1. We call the C*-envelope of $\ell^1(G, A)$ the C*-crossed product of A by G with respect to the action α and write as $A \rtimes_\alpha G$.

Let δ_s be the unitary element of $\ell^1(G, A)$ such that $\delta_s(s) = 1$

and $\delta_s(t) = 0$ if $s \neq t$. We have used the same notation in Sec. 3.1 for scalar cases but there should be no confusion for their definitions. Note that in the present case δ_s belongs to $A \rtimes_\alpha G$ and $\delta_s a \delta_s^* = \alpha_s(a)$ $(a \in A)$. Thus by using these unitaries a function $x = (x(s))$ in $k(G, A)$ is written as an expansion $x = \sum_s x(s)\delta_s$. Then operations in the algebra $k(G, A)$ simply mean the rules of changes of coefficients in the expansion. Namely, we have

$$(\Sigma x(s)\delta_s)^* = \Sigma \delta_s^* x(s)^* = \Sigma \delta_{s^{-1}} x(s)^*$$

$$= \Sigma \alpha_{s^{-1}}(x(s)^*)\delta_{s^{-1}} = \Sigma \alpha_s(x(s^{-1})^*)\delta_s \quad ,$$

$$(\Sigma x(s)\delta_s)(\Sigma y(s)\delta_s) = \sum_{s,t} x(s)\delta_s y(t)\delta_t$$

$$= \sum_{s,t} x(s)\alpha_s(y(t))\delta_{st}$$

$$= \sum_t (\sum_s x(s)\alpha_s(y(s^{-1}t)))\delta_t \quad .$$

A pair (π, u) of a representation π of A and a unitary representation u of G on the same space H is called a covariant representation of the C*-dynamical system $\{A, G, \alpha\}$ if $u_s \pi(a) u_s^* = \pi(\alpha_s(a))$. If we assume $A \rtimes_\alpha G$ is acting on the universal representation space H_u, the pair of the identity representation of A and $\delta_s (s \in G)$ is a covariant representation of $\{A, G, \alpha\}$.

<u>Theorem 3.2.1.</u> Every representation of $A \rtimes_\alpha G$ induces a covariant representation of $\{A, G, \alpha\}$. Conversely every covariant representation determines a representation of $A \rtimes_\alpha G$.

<u>Proof.</u> It is enough to show the second assertion. Let $\{\pi, u\}$ be a covariant representation of $\{A, G, \alpha\}$ on a Hilbert space H. Take x in $k(G, A)$ and define an operator $\tilde{\pi}(x)$ on H by

$$\tilde{\pi}(x) = \sum_s \pi(x(s))u_s \quad .$$

Then, similarly as calculated before for $x = \sum_s x(s)\delta_s$, we obtain that $\tilde{\pi}(x^*) = \tilde{\pi}(x)^*$ and for x, y in $k(G, A)$

$$\begin{aligned}\tilde{\pi}(xy) &= \sum_t \pi(xy(t))u_t \\ &= \sum_t \sum_s \pi(x(s))\pi(\alpha_s(y(s^{-1}t)))u_t \\ &= (\sum_s \pi(x(s))u_s)(\sum_t \pi(y(t))u_t) = \tilde{\pi}(x)\tilde{\pi}(y) \quad .\end{aligned}$$

Since $\tilde{\pi}$ is obviously norm decreasing, it extends to a representation of $\ell^1(G, A)$, hence to that of $A \rtimes_\alpha G$.

We write the above extended representation $\tilde{\pi}$ as $\pi \times u$.

Next we shall define the reduced crossed product $A \rtimes_{\alpha r} G$, which is more accessible than $A \rtimes_\alpha G$. Assume first that A is acting on a Hilbert space H and put $K = \ell^2(G) \otimes H$ which is also regarded as the H-valued ℓ^2-space on G, $\ell^2(G, H)$. Define a representation of A as well as a unitary representation of G on K by

$$(\pi_\alpha(a)\xi)(s) = \alpha_{s^{-1}}(a)\xi(s) \qquad \xi \in K \quad , \quad a \in A \quad ,$$

$$(\lambda_s\xi)(t) = \xi(s^{-1}t) \quad .$$

We then see that $\{\pi_\alpha, \lambda_s\}$ is a covariant representation of $\{A, G, \alpha\}$ and moreover π_α is a *-isomorphism.

<u>Definition 3.2.2.</u> The reduced C*-crossed product is the C*-algebra on K generated by the family of operators $\{\pi_\alpha(a), \lambda_s | a \in A, s \in G\}$. We denote it by $A \rtimes_{\alpha r} G$.

It can be proved that this definition does not depend on the acting space H of A. Actually, we have the following result.

Proposition 3.2.2. Let A and B be C*-algebras with actions α, β of G on them, respectively. Suppose there exists a *-isomorphism θ of A to B such that

$$\beta_s \cdot \theta(a) = \theta \cdot \alpha_s(a) \qquad a \in A \quad , \quad s \in G \quad ,$$

then there exists a *-isomorphism of $A \underset{\alpha r}{\rtimes} G$ to $B \underset{\beta r}{\rtimes} G$ which brings $\pi_\alpha(a)$ to $\pi_\beta(\theta(a))$ and λ_s^α to λ_s^β.

In the very simple case where $A = C$ with the trivial action of G, $\ell^1(G, A)$ coincides with $\ell^1(G)$ and $\{\lambda_s\}$ is the left regular representation of G on $\ell^2(G)$. Thus, in this case $A \underset{\alpha}{\rtimes} G$ is nothing but the group C*-algebra C*(G) and $A \underset{\alpha r}{\rtimes} G$ is the reduced group algebra $C_r^*(G)$ in Sec. 3.1. Next we consider the case where the action is trivial and A is a C*-algebra. Then, on the space $K = \ell^2(G) \otimes H$ we have

$$\pi_\alpha(a) = 1 \otimes a \quad a \in A \quad ; \quad \lambda_s = u_s \otimes 1 \quad s \in G \quad ,$$

where $s \to u_s$ is the usual left regular representation of G on $\ell^2(G)$. Therefore, by definition, $A \underset{\alpha r}{\rtimes} G$ coincides with the C*-tensor product $C_r^*(G)_{\min} A$ (C*-tensor product with the minimum C*-cross norm. cf. Appendix C). On the other hand, the algebra $\ell^1(G, A)$ is this time endowed with the usual convolution and, as its C*-envelope, the algebra $A \underset{\alpha}{\rtimes} G$ is known to coincide with the tensor product $C^*(G) \underset{\max}{\otimes} A$ (C*-tensor product with the maximum C*-cross norm). In this circumstance it is known that if G is amenable, not only is C*(G) canonically isomorphic to $C_r^*(G)$ but also C*-cross norm on the algebraic tensor product $C^*(G) \odot A$ is unique. This is a special case of the next deep:

Theorem 3.2.3. If G is amenable the canonical representation $\pi_\alpha \times \lambda$ of $A \underset{\alpha}{\rtimes} G$ to $A \underset{\alpha r}{\rtimes} G$ in (3.2.1) is faithful.

Therefore, if G is amenable we can take advantage of both

structures of $A \underset{\alpha}{\rtimes} G$ and $A \underset{\alpha r}{\rtimes} G$.

We consider next more detailed structure of $A \underset{\alpha r}{\rtimes} G$. Throughout this argument infinite sums of operators are understood with respect to the strong operator topology. Let w_s be an operator of K to H such that

$$w_s\xi = \xi(s^{-1}) \qquad \xi \in K \quad .$$

Then, $w_s^* w_s$ is the projection on the subspace $\{\xi \in K| \text{ only } \xi(s^{-1}) = 0\}$ with $w_s w_s^* = 1$ and $\{w_s^* w_s\}$ is a family of orthogonal projections with sum 1. Furthermore, the following basic calculation rules are easily verified;

$$w_s \lambda_t = w_{st} \quad , \qquad w_s \pi_\alpha(a) w_s^* = \alpha_s(a)$$

$$\sum_s w_s^* \alpha_s(a) w_s = \pi_\alpha(a) \quad .$$

Put $\varepsilon(x) = w_e x w_e^*$ for $x \in B(K)$. By definition, ε is a normal positive map of $B(K)$ to $B(H)$. We still denote the restriction of ε to $A \underset{\alpha r}{\rtimes} G$ by ε. Take an element $x_0 = \sum_s \pi_\alpha(x_0(s))\lambda_s$ for a function $x_0(s)$ in $k(G, A)$. We then see that $\varepsilon(x_0) = x_0(e)$. Therefore, $\varepsilon(A \underset{\alpha r}{\rtimes} G) = A$ and identifying A with $\pi_\alpha(A)$ the map ε is regarded as a (normal positive) projection of norm one from $A \underset{\alpha r}{\rtimes} G$ to A.

Next let x be an arbitrary element of $A \underset{\alpha r}{\rtimes} G$ and note first that

$$\varepsilon(\lambda_s x \lambda_s^*) = \alpha_s \cdot \varepsilon(x)$$

as the identity holds for dense elements with finite expansion. Put

$$x(s) = \varepsilon(x\lambda_s^*) \quad .$$

Then, $\| x(s) \| \le \| x \|$ and for every $s, t \in G$

$$w_s x w_t^* = w_e \lambda_s x \lambda_t^* w_e^* = \varepsilon(\lambda_s x \lambda^*_{s^{-1}t} \lambda_s^*)$$

$$= \alpha_s \cdot \varepsilon(x \lambda^*_{s^{-1}t}) = \alpha_s(x(s^{-1}t)) \quad .$$

Hence if $x(s) = 0$ for every s then $w_s x w_t^* = 0$ for every s, t and $x = 0$. Namely x is uniquely determined by the function $x(s)$ or by the coefficient set $\{x(s)\}$. We express this correspondence as

$$x \sim \{x(s)\} \qquad \text{or} \qquad x \sim \sum_s x(s)\lambda_s \quad ,$$

and call $\{x(s)\}$ the Fourier coefficients of x. Indeed, when specialized as $A = \mathbb{C}$, $G = Z$ with trivial action they are exactly Fourier coefficients of those functions in $C^*(Z) = C(T)$. Thus as a natural counterpart the above correspondence does not mean in general an actual expansion of x as an infinite series $\sum_s \pi_\alpha(x(s))\lambda_s$, for instance, in the strong topology. In this correspondence, however, if $x \sim \{x(s)\}$ and $y \sim \{y(s)\}$ we have

$$x^* \sim \{\alpha_s(x(s^{-1})^*)\}$$

$$xy \sim \{\sum_s x(s)\alpha_s(y(s^{-1}t))\} \quad ,$$

where the second correspondence is given by the calculation:

$$xy(t) = \varepsilon(xy\lambda_t^*) = w_e xy \lambda_t^* w_e^*$$

$$= \sum_s w_e x w_s^* w_s y w_t^* = \sum_s x(s)\alpha_s(y(s^{-1}t)) \quad .$$

Notice that one may calculate them as if $x = \Sigma x(s)\lambda_s$ and $y = \Sigma y(s)\lambda_s$.

Now let $\tilde{\pi}_r = \pi_\alpha \times \lambda$ be the canonical representation of $A \rtimes_\alpha G$ to $A \rtimes_{\alpha r} G$. We call a state τ on a C*-algebra B a tracial state if $\tau(xy) = \tau(yx)$ for $x, y \in B$ or equivalently $\tau(x^*x) = \tau(xx^*)$ for every $x \in B$ (cf. Appendix B).

Proposition 3.2.4.

(1) The map E on $\ell^1(G, A)$ extends to a positive projection of norm one of $A \rtimes_\alpha G$ to A with the property $\tilde{\pi}_\gamma \cdot E = \varepsilon \cdot \tilde{\pi}_\gamma$.

(2) ε is a faithful positive projection of norm one of $A \rtimes_{\alpha r} G$;

(3) If φ is an α-invariant tracial state on A, then $\hat{\tau} = \varphi \cdot E$ and $\tau = \varphi \cdot \varepsilon$ are tracial states on $A \rtimes_\alpha G$ and $A \rtimes_{\alpha r} G$ respectively for which τ is faithful if φ is faithful.

Proof.

(1) By definition of E and ε, we have

$$\tilde{\pi}_\gamma \cdot E(x) = \varepsilon \cdot \tilde{\pi}_r(x) \qquad \text{for} \qquad x \in \ell^1(G, A) \quad .$$

Since $\tilde{\pi}_r$ is isometric on A, it follows that

$$\| E(x) \| = \| \varepsilon \cdot \tilde{\pi}_r(x) \| \leq \| \tilde{\pi}_r(x) \| \leq \| x \|_\infty$$

for the C*-enveloping norm $\| x \|_\infty$ of x. Hence E extends to $A \rtimes_\alpha G$ with the same property for $\tilde{\pi}_\gamma$ and ε.

(2) We note that

$$\varepsilon(x^*x) = x^*x(e) = \sum_{s \in G} x^*(s)\alpha_s(x(s^{-1}))$$

$$= \sum_{s \in G} \alpha_s(x(s^{-1})^*x(s^{-1})) \quad .$$

Hence, $\varepsilon(x^*x) = 0$ implies that $x(s) = 0$ for every $s \in G$, and $x = 0$. The assertion (3) is clear by the values of E and ε on x^*x and xx^* together with (1).

We remark that the projection E is not faithful unless the group G is amenable.

Now let $\Sigma = (X, \sigma)$ be a topological dynamical system and let α

be the *-automorphism of $C(X)$ induced by σ. We may regard the system $\{C(X), \alpha\}$ as an action of Z on $C(X)$. Then, since Z is amenable we can identify $C(X) \rtimes_{\alpha} Z$ with $C(X) \rtimes_{\alpha r} Z$. Moreover, assuming $C(X)$ as acting on a space H, the unitary operator λ_1 corresponding to 1 is written as $1 \otimes s$ for the shift unitary operator on $\ell^2(Z)$. Therefore,

$$C(X) \rtimes_{\alpha r} Z = C^*(\pi_{\alpha}(C(X)), 1 \otimes s) \quad .$$

Suppose further that α is spatial on $C(X)$, that is, there is a unitary operator u on H such that $\alpha(f) = ufu^*$ for $f \in C(X)$. Then the unitary operator w on $\ell^2(Z, H)$ defined by

$$w(\xi)(n) = u^{-n}\xi(n) \qquad \text{for} \qquad \xi = (\xi(n))$$

intertwines the crossed product $C^*(\pi_{\alpha}(C(X)), 1 \otimes s)$ and the C*-algebra $C^*(C(X) \otimes 1, u \otimes s)$, so that the latter is also considered as the crossed product $C(X) \rtimes_{\alpha r} Z$. Indeed, one may easily verify that

$$w^{-1}\pi_{\alpha}(f)w = f \otimes 1 \quad f \in C(X) \quad ; \quad w^{-1}(1 \otimes s)w = u \otimes s \quad .$$

We shall sometimes use this definition of the crossed product for actions of Z.

3.3 Positive Linear Functionals of Transformation Group C*-algebras

We assume in our context that $A = C(X)$ and a discrete group G is acting as a group of homeomorphisms of the compact space X. Let α be the action of G on $C(X)$. The algebra $C(X) \rtimes_{\alpha} G$ is considered as a transplantation of the dynamical system $\Sigma = (X, G, \sigma_s)$ into an algebraic frame of C*-algebras. Thus we call $C(X) \rtimes_{\alpha} G$ or $C(X) \rtimes_{\alpha r} G$ the transformation group C*-algebra. In this section we shall discuss the structure of the set of positive linear functionals on these crossed products with the problem of pure state extension of pure states on $C(X)$. We also investigate the existence of tracial states on $C(X) \rtimes_{\alpha} G$.

We must, however, first discuss structure of positive linear functionals on $A \rtimes_{\alpha} G$ for a general C*-algebra A and an action α.

Let φ be a positive linear functional on $A \rtimes_{\alpha} G$ and define the function $\Phi\colon G \to A^*$ by

$$\Phi(s)(a) = \varphi(a\delta_s) \qquad s \in G \quad , \quad a \in A \quad .$$

The function Φ is called a positive definite function (with respect to the action α). In fact, if A reduces to the scalar this definition coincides with the usual one. Therefore as in (3.1.2) there is bijective correspondence between the set of positive linear functionals on $A \rtimes_{\alpha} G$ and that of positive definite functions on G denoted by PD(G).

<u>Proposition 3.3.1.</u> For a bounded function Φ from G to A^* the following conditions are equivalent;

(1) Φ is positive definite;

(2) $\sum_{i,j} \Phi(s_i^{-1}s_j)(\alpha_{s_{\pm}^{-1}}(a_i^* a_j)) \geq 0$ for all finite sets $\{s_i\}$ in G and $\{a_i\}$ in A.

<u>Proof.</u> (1)$\Rightarrow$(2). For a given finite sets $\{s_i\}$ and $\{a_i\}$, define the element $x = \sum_i a_i \delta_{s_i}$. Then $x^* = \sum_i \alpha_{s_i^{-1}}(a_i^*)\delta_{s_i^{-1}}$. Hence if φ is a positive linear functional on $A \rtimes_{\alpha} G$ corresponding to Φ, we have

$$0 \leq \varphi(x^*x) = \varphi\Big(\sum_{i,j} \alpha_{s_i^{-1}}(a_i^* a_j)\delta_{s_i^{-1}s_j} \Big)$$

$$= \sum_{i,j} \Phi(s_i^{-1}s_j)(\alpha_{s_i^{-1}}(a_i^* a_j)) \quad .$$

(2)$\Rightarrow$(1). Define the functional φ on $k(G, A)$ by

$$\varphi(x) = \sum_i \Phi(s_i)(a_i) \qquad \text{for} \qquad x = \sum_i a_i\delta_{s_i} \quad .$$

Since Φ is bounded, φ is clearly continuous in ℓ_1-norm and the above calculation shows that $\varphi(x^*x) \geq 0$. It follows that φ extends to a positive linear functional on $\ell^1(G, A)$, and finally to $A \rtimes_\alpha G$. One may then verify easily that the positive definite function associated with φ coincides with Φ.

We say that the function Φ is pure if φ is pure.

Now coming back to the original situation, we recall that $C(X)$ is regarded as an abelian C*-subalgebra of $C(X) \rtimes_\alpha G$ (we hereafter abbreviate this algebra by A_Σ for a system $\Sigma = (X, G, \sigma_s)$ sometimes) consisting of those functions f's on G whose values on the unit e are f's and 0 elsewhere. In the following we write the action of G on X simply as sx instead of $\sigma_s x$.

In order to consider pure state extension of the pure states of $C(X)$ we have to first check when $C(X)$ is maximal abelian in A_Σ. For if $C(X)$ is not a maximal abelian C*-subalgebra of A_Σ pure state extensions from $C(X)$ to a commutative C*-subalgebra B of A_Σ which contains $C(X)$ strictly are already not unique in general by the Stone-Weierstrass theorem. Let φ be a state on A_Σ and write $\varphi = \sum_s \oplus \mu_s$ where $s \to \mu_s$ is the corresponding positive definite function on G regarded as a distribution of measures on X. Put $X_s = \{x \in X \mid sx = x\}$ for $s \in G$.

<u>Proposition 3.3.2.</u> Suppose that G is amenable. Then $C(X)$ is a maximal abelian C*-subalgebra of A_Σ if and only if every set X_s has no interior except X_e (= X).

<u>Proof.</u> Since we can identify A_Σ with $C(X) \rtimes_{\alpha, r} G$, an element a of A_Σ is expressed as a formal sum $\sum_s f_s\delta_s$ with Fourier coefficients $\{f_s\}$. In this connection, it is enough to show that

$$C(X)' \cap A_\Sigma = \{a = \sum_s f_s\delta_s \mid \text{supp } f_s \subset X_{s^{-1}}, s \in G\} \quad .$$

The above equality follows from the following successive equivalent assertions:

$$(\sum_s f_s\delta_s)g = g(\sum_s f_s\delta_s) \qquad g \in C(X) \quad ,$$

$$\Leftrightarrow \Sigma f_s\alpha_s(g)\delta_s = \Sigma g f_s\delta_s$$

$$\Leftrightarrow f_s\alpha_s(g) = gf_s = f_s g \qquad g \in C(X) \quad , \quad s \in G$$

$$\Leftrightarrow f_s(x)g(s^{-1}x) = f_s(x)g(x) \qquad x \in X \quad ,$$

$$\Leftrightarrow \text{supp } f_s \subset X_s \qquad s \in G \quad .$$

Corollary 3.3.3. If $G = Z$, $C(X)$ is a maximal abelian C*-subalgebra of A_Σ if and only if for every positive integer n the set of points in X with period n has no interior.

On the other hand, with the assumption $G = Z$ if all points of X are periodic $C(X)$ is not maximal abelian in A_Σ. At this point we show the next

Example 3.3.1. Let X be a countable union of circles in the complex plane with the same center 0, whose radii converge increasingly to 1 together with the circle of radius 1. Let X_n be the n-th circle and write X_0 the last circle of radius 1. We consider the homeomorphism in $X = \bigcup_{n=0}^{\infty} X_n$ which induces the rotation of the angle $\frac{2\pi}{n}$ on X_n and the identity map on X_0. Then all points in X are periodic points (including fixed points) and for every n there is a point of period n.

Now consider a state $\varphi = \sum_s \oplus \mu_s$ on A_Σ. As a functional on $C(X)$ each μ_s $(s \neq e)$ has the decomposition

$$\mu_s = \mu_s^1 + i\mu_s^2$$

into its real part and imaginary part where

$$\mu_s^1(f) = \frac{1}{2}\left(\mu_s(f) + \overline{\mu_s(\bar{f})}\right)$$

$$\mu_s^2(f) = \frac{1}{2i}\left(\mu_s(f) - \overline{\mu_s(\bar{f})}\right) \qquad f \in C(X) \quad .$$

Here the measures μ_s^1 and μ_s^2 are real signed ones on X. Define the positive measure $|\mu_s|$ as

$$|\mu_s| = |\mu_s^1| + |\mu_s^2| \quad .$$

We say that μ_s is absolutely continuous with respect to μ_e (probability measure), $\mu_s < \mu_e$ if both μ_s^1 and μ_s^2 are absolutely continuous with respect to μ_e. This, in turn, is equivalent to say that $|\mu_s^1| < \mu_e$ and $|\mu_s^2| < \mu_e$, hence $|\mu_s| < \mu_e$.

Proposition 3.3.4. With above notations for a state φ on A_Σ, we have

(1) $\mu_s < \mu_e$ and $\mu_s < \mu_e \cdot \alpha_s^{-1}$;

(2) $\mu_{s^{-1}}(f) = \overline{\mu_s(\alpha_s(\bar{f}))} \qquad f \in C(X) \quad , \quad s \in G \quad .$

Proof. Let E be a Borel set of X. We assert that $\mu_e(E) = 0$ implies $\mu_s(E) = 0$, i.e. $\mu_s^1(E) = \mu_s^2(E) = 0$. Indeed, by the regularity of μ_e and $|\mu_s|$ there exist a compact set C and an open set U such that

$$C \subset E \subset U \quad \text{and}$$

$$\mu_e(U - C) < \varepsilon \qquad\qquad |\mu_s|(U - C) < \varepsilon \quad .$$

Take a continuous function f on X such that $f(x) = 1$ on C, $f(x) = 0$ on U^c and $0 \le f(x) \le 1$. Then,

$$\mu_e(f^2) = \int_C f^2 d\mu_e + \int_{U-C} f^2 d\mu_e$$

$$\le \mu_e(C) + \mu_e(U - C) < \varepsilon \quad ,$$

and

$$|\mu_s(f)| = |\int_U f d\mu_s| \geq |\int_C f d\mu_s| - |\int_{U-C} f d\mu_s|$$

$$\geq |\mu_s(C)| - \varepsilon \quad .$$

Hence by (2.3.3)

$$|\mu_s(C)| \leq |\mu_s(f)| + \varepsilon = |\varphi(f\delta_s)| + \varepsilon$$

$$\leq \mu_e(f^2) + \varepsilon < 2\varepsilon \quad ,$$

and

$$|\mu_s(E)| \leq |\mu_s(C)| + |\mu_s|(U - C) < 3\varepsilon \quad .$$

Therefore, $\mu_s(E) = 0$. If we make use of another identity $\mu_s(f) = \varphi(\delta_s \alpha_{s^{-1}}(f))$, almost same argument applying (2.3.3) leads us to the other conclusion $\mu_s < \mu_e \cdot \alpha_{s^{-1}}$.

As for the assertion (2), it is enough to notice that by (3.3.1) the matrix $|\mu_{s_i^{-1}s_j}(\alpha_{s_i^{-1}}(\bar{f}_i f_j))|$ is positive, hence self-adjoint and

$$\mu_{s_i^{-1}s_j}(\alpha_{s_i^{-1}}(\bar{f}_i f_j)) = \mu_{s_j^{-1}s_i}(\alpha_{s_j^{-1}}(\bar{f}_j f_i)) \quad .$$

In fact, considering the finite sets $\{e, s^{-1}\}$ and $\{1, f\}$ we get the conclusion.

Let H be a subgroup of G. We shall show that there is a bijective correspondence between the set of positive definite functions on G with supports contained in H and the set of positive definite functions on H. This will be seen from the following

<u>Lemma 3.3.5.</u> A scalar valued positive definite function Φ on a subgroup H extends to a positive definite function $\hat{\Phi}$ on G with $\text{supp}\, \hat{\Phi} \subseteq H$.

<u>Proof.</u> Let $\hat{\Phi}$ be the function on G such that $\hat{\Phi}|H = \Phi$ and $\hat{\Phi}(s) = 0$ elsewhere. Let $S = \{s_i\}$ be a finite subset of G. Write

$S = \bigcup_{k=1}^{m} S_k$, where S_k consists of those elements in S in the same left coset of H. Let $\{s_{1k}, 2_{2k}, \ldots, s_{\ell k}\}$ be the elements of S_k, then we can write $s_{ik} = st_i$ for $t_i \in H$. Now let $\{\lambda_i\}$ be a finite set of complex numbers indexed following the elements of S. Then, for a fixed k,

$$\sum_{i,j} \hat{\Phi}(s_{ik}^{-1}s_{jk})\bar{\lambda}_{ik}\lambda_{jk} = \sum_{i,j} \Phi(t_i^{-1}t_j)\bar{\lambda}_{ik}\lambda_{jk} \geq 0 \quad .$$

On the other hand, if s_i and s_j are not in the same set S_k, $s_i^{-1}s_j \notin H$ and $\hat{\Phi}(s_i^{-1}s_j) = 0$. After all,

$$\sum_{i,j} \hat{\Phi}(s_i^{-1}s_j)\bar{\lambda}_i\lambda_j = \sum_{k=1}^{m} \sum_{i,j} \hat{\Phi}(s_{ik}^{-1}s_{jk})\bar{\lambda}_{ik}\lambda_{jk} \geq 0 \quad ,$$

hence by (3.1.2) $\hat{\Phi}$ is a positive definite function.

For an invariant positive measure μ there is a simple way to construct a positive linear functional on A_Σ by means of a positive definite function. Namely we have

<u>Proposition 3.3.6.</u> Let Φ be a scalar valued positive definite function on G, then the functional $\varphi = \sum_S \oplus \Phi(s)\mu$ on A_Σ is positive.

<u>Proof.</u> Let $\{s_i\}$ and $\{f_i\}$ be finite sets in G and $C(X)$ respectively. We have

$$\sum_{i,j} \Phi(s_i^{-1}s_j)\mu(\alpha_{s_i^{-1}}(\bar{f}_i f_j)) = \sum_{i,j} \Phi(s_i^{-1}s_j)\mu(\bar{f}_i f_j)$$

$$= |\Phi(s_i^{-1}s_j)| \cdot |\mu(\bar{f}_i f_j)| \geq 0 \quad ,$$

where the above product means the Schur product of matrices. The Schur product of positive matrices is again positive. Thus by (3.3.1) the function: $s \to \Phi(s)\mu$ is positive definite and φ is a positive functional.

Note that the method is not available for usual positive measures.

Let G_x be the isotropy subgroup for x, that is,

$$G_x = \{s \in G \mid sx = x\} \quad .$$

Theorem 3.3.7. Let μ_x be a pure state of $C(X)$ associated to a point x in X. We have then;

(1) There exists a bijective correspondence between the set of state extensions of μ_x to A_Σ and that of positive definite functions $\check{\Phi}$ on G_x with $\check{\Phi}(e) = 1$. The correspondence is given by $\varphi = \Sigma \oplus \Phi(s)\mu_x \leftrightarrow \check{\Phi} = \Phi | G_x$.

(2) The state extension φ is pure if and only if the corresponding function $\check{\Phi}$ is pure. The pure state μ_x has a unique pure state extension if and only if $G_x = \{e\}$.

Proof. (1) Let μ_s be the positive definite function on G associated with a state extension φ of μ_x. By (3.3.4)

$$\operatorname{supp}|\mu_s| \subseteq \operatorname{supp} \mu_e = \operatorname{supp} \mu_x = \{x\} \quad ,$$

hence $\mu_s = \Phi(s)\mu_x$ for a scalar $\Phi(s)$. On the other hand, we also have the inclusion

$$\operatorname{supp}|\mu_s| \subseteq \operatorname{supp} \mu_x \circ \alpha_{s^{-1}} \quad .$$

Hence if $sx \neq x$, $\mu_s = 0$. Namely, Φ is supported in G_x. Let $\{s_i\}$ and $\{f_i\}$ be finite sets in G_x and $C(X)$ respectively. Then

$$\begin{aligned} 0 &\leq \sum_{i,j} \mu(s_i^{-1}s_j)(\alpha_{s_i^{-1}}(\bar{f}_i f_j)) \\ &= \sum_{i,j} \Phi(s_i^{-1}s_j)\bar{f}_i(s_i x) f_j(s_i x) \\ &= \sum_{i,j} \Phi(s_i^{-1}s_j)\bar{f}_i(s_i x) f_j(s_j x) \\ &= [\Phi(s_i^{-1}s_j)] \cdot [\bar{f}_i(s_i x) f_j(s_j x)] \quad \text{(Schur product)} \quad . \end{aligned}$$

Since the $\{f_i\}$ is arbitrary, this means that the matrix $[\Phi(s_i^{-1}s_j)]$ is positive. Therefore by (3.1.2) the function $\check{\Phi} = \Phi|G_x$ is positive definite on G_x.

Conversely let $\check{\Phi}$ be a positive definite function on G_x with $\check{\Phi}(e) = 1$ and let Φ be the extended positive definite function on G by (3.3.5). We define the bounded linear functional φ on $k(G, C(X))$ by

$$\varphi(\sum_i f_i \delta_{s_i}) = \sum_i \Phi(s_i) f_i(x) \quad .$$

Then for an element $a = \sum_i f_i \delta_{s_i}$ in $k(G, C(X))$

$$\varphi(a^*a) = \sum_{i,j} \Phi(s_i^{-1}s_j)\bar{f}_i(s_i x) f_j(s_i x)$$

$$= \Sigma_1 \check{\Phi}(s_i^{-1}s_j) f_i(s_i x) f_j(s_i x) \quad ,$$

where the sum Σ_1 is ranging over G_x, that is, over only those elements such that $s_i^{-1}s_j \in G_x$. This equals to

$$\sum_{i,j} \Phi(s_i^{-1}s_j)\bar{f}_i(s_i x) f_j(s_j x)$$

$$= [\Phi(s_i^{-1}s_j)] \cdot [\bar{f}_i(s_i x) f_j(s_j x)] \geq 0 \quad .$$

The last inequality follows from the fact that both matrices are positive definite. Therefore φ extends to a state on A_Σ and moreover

$$\varphi(f) = \check{\Phi}(e) f(x) = \mu_x(f) \qquad f \in C(X) \quad .$$

For the assertion (2) one may easily verify the first part together with one implication of the second part. To show the other implication we note that if the pure state extension of μ_x is unique its state extension is also unique, for the set of state extensions is the closed convex hull of its extreme points, pure state extensions. Therefore in this case, by (1) the group C*-algebra $C^*(G_x)$ becomes one dimensional. Hence G_x must be the trivial subgroup. This completes all proofs.

In the last step of the proof one may also finish the argument specifying different positive definite functions on G_x when $G_x \neq \{e\}$, such as the constant positive definite function Φ_1 with value 1 and the other Φ_2 with $\Phi_2(e) = 1$ and 0 elsewhere.

From this theorem, if $G_x = \{e\}$ for every x in X every pure state on C(X) has a unique pure state extension to A_Σ. When $\Sigma = (X, \sigma)$, this is equivalent to say that there are no periodic points for σ. The next example which is closely related to our discussion in the last chapter shows such a case.

Example 3.3.2. Let βZ be the Stone-Čech compactification of the integer group Z. We consider in βZ the homeomorphism σ which is the extension of the shift on Z, that is, $\sigma(n) = n + 1$. We shall show that for a fixed positive integer k there are no k-periodic points. In fact, suppose that there was a k-periodic point ω in $\beta Z \sim Z$. Take a bounded function f on Z such that

$$f(n(k+1)+j) = j \qquad 0 \leq j \leq k \quad , \qquad n \in Z \quad .$$

Let $\{n_\alpha\}$ be a net of Z converging to ω. Then, regarding f as a continuous function on βZ, we have

$$f(n_\alpha) \to f(\omega) \qquad \text{and}$$

$$f(n_\alpha + k) = f(\sigma^k(n_\alpha)) \qquad f(\sigma^k(\omega)) = f(\omega) \quad ,$$

whereas by definition of f

$$|f(n_\alpha) - f(n_\alpha + k)| \geq 1 \qquad \text{for every } n_\alpha \quad .$$

This is a contradiction.

We remark that this dynamical system is usually treated in the form (Z, σ) as the simplest example of measure theoretical dynamical system and hardly appears in the present form in a standard context of topological dynamical systems. However problems about states and pure states on $C(\beta Z)$ are the problems in the category of C*-algebras and not in the category of von Neumann algebras. As mentioned before we shall discuss later interesting examples of topological dynamical systems closely related to the above example.

Let φ be a pure state extension of μ_x and let $\{H_\varphi, \pi_\varphi, \xi_\varphi\}$ be the GNS-representation of φ, which induces a covariant representation of $\{C(X), G, \alpha\}$. We shall determine the structure of $\{\pi_\varphi, H_\varphi\}$. Write $u_s = \pi_\varphi(\delta_s)$. We define the subspace

$$H_s = \{\xi \in H \mid \pi_\varphi(f)\xi = f(sx)\xi \quad f \in C(X)\} \qquad .$$

The subspace is determined modulo the left coset of G_x and one may easily verify that $\xi \in H_e$. We assert that if $s^{-1}t \notin G_x$ the subspaces H_s and H_t are orthogonal. In fact, we may choose a function $f \in C(X)$ such that $f(sx) = 1$ and $f(tx) = 0$ so that for every vector $\xi \in H_s$ and $\eta \in H_t$,

$$\begin{aligned}(\xi, \eta) &= (f(sx)\xi, \eta) = (\pi_\varphi(f)\xi, \eta)\\ &= (\xi, \pi_\varphi(\bar{f})\eta) = 0 \qquad .\end{aligned}$$

On the other hand, $u_s H_t = H_{st}$ and the subspace $\sum_{s_\alpha} \oplus H_{s_\alpha}$ ranging over the representatives $\{s_\alpha\}$ of the left cosets of G_x is invariant under $\pi_\varphi(A_\Sigma)$. Hence, $H_\varphi = \sum_{s_\alpha} \oplus H_{s_\alpha}$. Here the isotropy subgroup G_x acts irreducibly on the subspace H_e. For if K_0 were a non-trivial invariant subspace of H_e under $\pi_\varphi(G_x)$ the subspace $K = \sum_{s_\alpha} \oplus u_{s_\alpha} K_0$ became a non-trivial subspace for $\pi_\varphi(A_\Sigma)$. Now for an element $s \in G$ write $ss_\alpha = s_\beta t$ for some $t \in G_x$. Since $H_{s_\alpha} = u_{s_\alpha} H_e$ and $H_{s_\beta} = u_{s_\beta} H_e$, the operator u_s acts on a vector $u_{s_\alpha}\xi$ for $\xi \in H_e$ as

$$u_{s_\alpha}\xi \to \xi \to u_t\xi \to u_{s_\beta}u_t\xi \quad .$$

Thus in this way the representation π_φ is induced by an irreducible representation of G_x and a one dimensional representation of $C(X)$ on H_e. Therefore, from the point of view of irreducible representations of A, abundance of pure state extensions of μ_x is based on that of irreducible representations of G_x on the space H_e. This provides another evidence of the assertion (2) of Theorem 3.3.7. The representation of G on H is also regarded as the one induced by the representation of G_x on H_e.

We shall precisely define this type of representation as an induced covariant representation of $\{C(X), G, \alpha\}$ in the next chapter and discuss in detail irreducible such representations of A_Σ.

Though our treatment will be somewhat different from standard context of induced representations of transformation group C*-algebras and those of locally compact groups even if G is discrete, it is usually an important problem for transformation group C*-algebras (resp. for locally compact groups) whether or not an irreducible representation or a representation of them is induced by a covariant representation of the algebra and an isotropy subgroup (resp. induced by a representation of a subgroup). Actually there was a long standing conjecture by Effros-Hahn (cf. Ref. 6) until the work by Gootman and Rosenberg[12]: In the general setting of the transformation group C*-algebras $C_0(X) \rtimes_\alpha G$ where $C_0(X)$ is the algebra of all continuous functions on a second countable locally compact space X vanishing at infinity and G is a second countable locally compact amenable group acting on X, whether or not every primitive ideal of $C_0(X) \rtimes_\alpha G$ (a closed ideal which is the kernel of an irreducible representation of $C_0(X) \rtimes_\alpha G$) becomes the kernel of some irreducible representation induced by a representation of an isotropy subgroup of G and that of $C_0(X)$. There are extensive literature about induced representations of locally compact groups and induced covariant representations under suitable countability conditions as well as about the conjecture and its generalized form. Our discussion in the next chapter is closely related to this thing, whereas we impose no countability condition for groups and spaces.

Next we shall discuss existence of tracial states on A_Σ.

<u>Proposition 3.3.8.</u> Let μ be a G-invariant probability measure in X and φ be a state extension of μ to A_Σ with the distribution of measures $\varphi = \Sigma \oplus \mu_s$. Then φ is a tracial state if and only if the following two conditions hold:

(1) $\operatorname{supp} \mu_s \subset X_s$; (cf. (3.3.2))

(2) $\mu_{s^{-1}ts}(f) = \mu_t(\alpha_s(f))$.

<u>Proof.</u> The assertion that φ is a tracial state is, by definition, equivalent to the following successive identities:

$$\varphi(f\delta_t g\delta_s) = \varphi(g\delta_s f\delta_t) \qquad f, g \in C(X) \quad , \quad s, t \in G \quad ,$$

$$\Longleftrightarrow \varphi(f\alpha_t(g)\delta_{ts}) = \varphi(g\alpha_s(f)\delta_{st})$$

$$\Longleftrightarrow \mu_{ts}(f\alpha_t(g)) = \mu_{st}(g\alpha_s(f)) \tag{3.1}$$

Therefore,

$$\mu_{s^{-1}ts}(f) = \mu_{s \cdot s^{-1}t}(\alpha_s(f)) = \mu_t(\alpha_s(f))$$

and

$$\mu_s(g\alpha_s(f)) = \mu_s(fg) \quad .$$

Take a point $x \notin X_s$. Then there exists a neighborhood $U(x)$ such that $U(x) \cap sU(x) = \phi$. Thus if we take a function f with $\operatorname{supp} f \subseteq U(x)$, we have by the above identity,

$$\mu_s(f^2) = \mu_s(f\alpha_s(f)) = 0 \quad .$$

This implies that $\mu_s(f) = 0$ for a positive function f and finally $\mu_s(f) = 0$ whenever $\operatorname{supp} f \subseteq U(x)$. Therefore, $x \notin \operatorname{supp} \mu_s$ and supp $\mu_s \subseteq X_s$. For other implication we shall show that two conditions imply (3.1). In fact,

$$\mu_{ts}(f\alpha_t(g)) = \mu_{s^{-1}\cdot sts}(f\alpha_t(g))$$

$$= \mu_{st}(\alpha_s(f\alpha_t(g)))$$

$$= \int_{X_{st}} \alpha_s(f)(x)\alpha_{st}(g)(x)d\mu_{st}$$

$$= \int_{X_{st}} f(s^{-1}x)g(x)d\mu_{st}$$

$$= \mu_{st}(\alpha_s(f)g) = \mu_{st}(g\alpha_s(f)) \quad .$$

This completes the proof.

When an invariant probability measure μ on X is given a tracial state extension τ of μ to A_Σ is always obtained by setting $\tau = \Sigma \oplus \mu_s$ with $\mu_e = \mu$ and $\mu_s = 0$ else. Namely, $\tau = \mu \cdot E$ where E is the canonical projection of norm one of A_Σ to $C(X)$. As a consequence of the above proposition we shall answer the question when μ has a unique tracial state extension provided that G is abelian.

Proposition 3.3.9. Let μ be an invariant probability measure in X and suppose that G is abelian. Then, μ has a unique tracial state extension if and only if $\mu(X_s) = 0$ for every s except the unit e.

Proof. Let $\tau = \Sigma \oplus \mu_s$ be a tracial state extension of μ. If $\mu(X_s) = 0$ for $s \neq e$, then $\mu_s = 0$ for such s and τ coincides necessarily with $\mu\ E$, the canonical tracial state extension of μ. Suppose next that there exists an element $t \neq e$ of G with $\mu(X_t) \neq 0$. We shall construct another tracial state extension of μ. Put $\mu_e = \mu$ and

$$\mu_t = \mu_{t^{-1}} = \frac{1}{2}\mu|X_t \quad .$$

Define the bounded linear functional τ on $\ell^1(G, C(X))$ by the distribution of measures:

$$\tau = \mu_e \oplus \mu_t \oplus \mu_{t^{-1}} .$$

Since,

$$sX_t = X_{sts^{-1}} = X_t \qquad s \in G ,$$

for every $f \in C(X)$ and $s \in G$ we have

$$\mu_t(\alpha_s(f)) = \mu_{t^{-1}}(\alpha_s(f)) = \mu_t(f) .$$

The functional τ satisfies the conditions in (3.3.8). We assert that τ is positive definite and hence it extends to a tracial state on A_Σ. In fact, take an element $a = \sum_s f_s\delta_s$ in $k(G, C(X))$. Then,

$$\begin{aligned}
\tau(a^*a) &= \tau(\sum_s \sum_u \alpha_{s^{-1}}(\bar{f}_s f_{su})\delta_u) \\
&= \mu(\sum_s \alpha_{s^{-1}}(\bar{f}_s f_s)) + \mu_t(\sum_s \alpha_{s^{-1}}(\bar{f}_s f_{st})) \\
&\qquad + \mu_{t^{-1}}(\sum_s \alpha_{s^{-1}}(\bar{f}_s f_{st^{-1}})) \\
&= \mu(\sum_s |f_s|^2) + \mu_t(\sum_s \bar{f}_s f_{st}) + \mu_{t^{-1}}(\sum_u \bar{f}_{ut} f_u) \\
&\geq \frac{1}{2} \sum_s \int_{X_t} (\bar{f}_s(x) f_s(x) + \bar{f}_s(x) f_{st}(x) \\
&\qquad + \bar{f}_{st}(x) f_s(x) + \bar{f}_{st}(x) f_{st}(x)) d\mu \\
&= \frac{1}{2} \sum_s \int_{X_t} |f_s(x) + f_{st}(x)|^2 d\mu \geq 0 .
\end{aligned}$$

This completes the proof.

If there is no fixed point for every $s \in G$ such as Example (3.3.2) and (1.1.2) every invariant measure on X has a unique tracial state extension. Besides, in Example 1.1.2 the invariant measure is also unique as the Lebesque measure. Consequently, in this example the crossed product $A_\theta = C(T) \rtimes_\sigma Z$ has a unique tracial state. This C*-algebra is called the irrational rotation algebra, which occupies the important rôle in recent development of the theory of operator algebras. We shall return to this example in the next chapter and furthermore in the last chapter. In connection with this example we state th next

Corollary 3.3.10. Keep the above assumption for G. Then the C*-algebra A_Σ has a unique tracial state if and only if the following conditions hold:

(1) $\Sigma = (X, G)$ is uniquely ergodic;

(2) For every $s \in G$ with $s \neq e$, $\mu(X_s) = 0$.

On the contrary, if $G = Z$ and if all points in X are periodic, then for any invariant probability measure there exists an integer n such that $\mu(X_n) = 0$, where X_n is the set of points with $\sigma^n x = x$. Therefore, in this case, tracial state extensions of invariant measures are always not unique.

APPENDIX C
CLASSES OF C*-ALGEBRAS

In this appendix, we do not assume that C*-algebras are unital and we describe about classes of C*-algebras. All results are stated without proofs.

As we already know, the simplest examples of C*-algebras are the algebra $C(X)$ and the algebra $C(H)$ the latter of which coincides with the matrix algebra M_n when H is finite dimensional. It is then natural to define the first class of C*-algebras called liminal algebras as the C*-algebras every image of whose irreducible representations coincides with the algebra of compact operators, $C(H)$. When a C*-algebra A is unital, A is liminal if and only if every irreducible representation is finite dimensional. Other simple example of liminal algebras are the C*-tensor product $C(X)\ M_n$ and direct sums of finite number of this type of algebras. The spatial C*-tensor products of C*-algebras are defined as follows. Let A and B be C*-algebras acting on the Hilbert spaces H and K respectively. Let $H \otimes K$ be the tensor product of Hilbert spaces whose inner product is defined as

$$(\xi_1 \otimes \eta_1, \xi_2 \otimes \eta_2) = (\xi_1, \xi_2)(\eta_1, \eta_2) \quad .$$

For $a \in A$ and $b \in B$ we consider the bounded linear operator $a \otimes b$ on $H. \otimes K$ defined by

$$a \otimes b(\xi \otimes \eta) = a\xi \otimes b\eta \quad .$$

The spatial tensor product $A \otimes B$ is then the C*-algebra generated by these operators. It is shown that the C*-algebra $A \otimes B$ does not depend on the choice of acting spaces.

C*-algebras which are, in a sense, piling of liminal algebras are called postliminal algebras. To be more precise, a postliminal algebra is defined as a C*-algebra every nonzero quotient C*-algebra of which possesses a non-zero closed liminal ideal. It is known that a C*-algebra A always contains a unique maximal postliminal ideal K such that the quotient algebra A/K has no nonzero postliminal ideal. If this ideal is trivial, A is called an antiliminal algebra. Every C*-subalgebra of a postliminal algebra as well as every quotient C*-algebra is also a postliminal algebra. On the other hand, a C*-algebra A is said to be of type I if every representation of A yields a von Neumann algebra of type I (cf. Appendix B).

Theorem 3.1. For a separable C*-algebra A, the following assertions are equivalent;

(1) A is postliminal;

(2) A is of type I;

(3) For each irreducible representation (π, H) of A, $\pi(A) \supset C(H)$;

(4) Two irreducible representations π_1 and π_2 of A are unitarily equivalent if and only if $\pi_1^{-1}(0) = \pi_2^{-1}$.

For a non-separable C*-algebra, whether or not we have the same equivalence assertions is still unknown. The problem is the equivalence of the assertion (4) and other group of assertions and it depends on the solution of the outstanding problem (Naimark's problem) about the characterization of the algebra $C(H)$ as the C*-algebra with unique dual, that is, unique irreducible representation within unitary equivalence. On the contrary, each separable C*-algebra which is not postliminal has factor representations both of type II and type III. Besides, such a C*-algebra has uncountably many pairwise disjoint representations, all with the same kernel. As in the case of C*-algebras we call a locally compact group G is of type I or postliminal if all its unitary representations are of type I. Then a semi-simple Lie group is of type I and a nilpotent Lie group is also known to be of type I.

A simple example of non-postliminal C*-algebra is a direct sum of $n \times n$ matrix algebras $A = \sum_{n=1}^{\infty} M_n$. In fact, it can be shown that if A is postliminal all but a finite number of component algebras must be matrix algebras with bounded degree. A notable class of anti-liminal algebras is the class of UHF (uniformly hyperfinite) algebras which is defined as the class of infinite C*-tensor products of matrix algebras. A UHF algebra is simple with unique tracial state. If we understand the concept of inductive limit of C*-algebras as being supplied by usual algebraic definition with the norm closure operation at its final stage, the class of UHF algebras apparently remains within the category of inductive limit of C*-algebras of type I. Furthermore, UHF algebras also belong to another important class of C*-algebras, called AF-algebras (approximately finite algebras). An AF-algebra is defined as a C*-algebra which is an inductive limit of a sequence of finite dimensional C*-algebras. A special feature of a UHF algebra is that it contains an increasing sequence of matrix algebras having the same unit as the whole algebra.

We shall introduce finally the class of nuclear C*-algebras. Let $A \odot B$ be the algebraic tensor product of two C*-algebras A and B, which becomes naturally a *-algebra. There are in general many C*-norms on $A \odot B$ which satisfy the cross norm property $\| a \bar{\otimes} b \| = \| a \| \| b \|$ for all $a \in A$ and $b \in B$. It is then known that there exist the largest C*-cross norm and the smallest C*-cross norm on $A \odot B$ the completion of which are denoted by $A \underset{max}{\otimes} B$ and $A \underset{min}{\otimes} B$ respectively. The latter coincides with spatial tensor product defined before. A C*-algebra A is then said to be nuclear if the C*-cross norm on $A \odot B$ is unique for any algebra B. A postliminal C*-algebra is nuclear. The reduced group C*-algebra $C_r^*(G)$ for a discrete group G is nuclear if and only if G is amenable. A special feature of this class is at the point that it is closed under the operation of inductive limit. Thus a UHF algebra is also a nuclear C*-algebra. On the other hand, the irrational rotation C*-algebra A_θ is nuclear as the crossed product of the commutative C*-algebra $C(T)$ by the amenable group Z, but it is not an AF-algebra. Nevertheless, it is further shown that A_θ can be imbedded into an AF-algebra. Hence a C*-subalgebra of an AF-algebra need not be an AF-algebra in general, whereas quotient C*-algebras are easily seen to be AF-algebras. Nuclear C*-algebras are also known to have the same type of hereditary properties. However it took considerable time to settle all of these questions since these classes were introduced. As stated before about A, the transformation group C*-algebra $C(X) \rtimes_\alpha G$ belongs to the class of nuclear C*-algebras if G is amenable but usually it is not an AF-algebra. Relations of the algebra $C(X) \rtimes_\alpha G$ to the class of liminal (postliminal) algebras will be discussed in the next chapter with respect to the properties of dynamical systems.

Nuclear C*-algebras are characterized in the following way.

Theorem 3.2. The following assertions for a C*-algebra A are equivalent;

(1) A is nuclear;

(2) Every representation of A gives rise to an injective von Neumann algebra.

An example of a nuclear C*-algebra which cannot be reached as an inductive limit of C*-algebras of type I, called the Cuntz algebra $O(n)$, is defined as the C*-algebra generated by those isometries $\{s_1, s_2, \ldots, s_n\}$ whose ranges are disjoint and $\sum_{i=1}^{n} s_i s_i^* = 1$. The algebra $O(n)$ does not depend on the choice of those isometries and it has many other interesting properties.

CHAPTER 4

ORBIT STRUCTURE AND DISCRETE IRREDUCIBLE REPRESENTATION OF TRANSFORMATION GROUP C*-ALGEBRAS

This chapter consists of four sections. In 4.1 we define discrete induced covariant representations of transformation group C*-algebras arised from isotropy subgroups of the acting group G and analyse their basic structure. The results will lead to many applications in the forthcoming discussion. The section 4.2 is concerning topological equivalence between the spaces of finite dimensional irreducible representations of the C*-algebra A_Σ for a dynamical system $\Sigma = (X, \sigma)$ and the sets of periodic points for Σ. In Sec. 4.3 we discuss relations between ideal structure of A_Σ and orbit structure of Σ. The last section is devoted to the analysis of representations of the 3-dimensional discrete Heisenberg group as applications of previous results.

4.1 Discrete Irreducible Representation of Transformation Group C*-algebras and Finite Orbits of Their Dynamical Systems

We keep the assumption that a discrete group G is acting effectively on a compact space X as a group of homeomorphisms. Write

$$X_n = \{x \in X \mid \operatorname{card}(o_G(x)) = n\} \quad ,$$

$$X^n = \{x \in X \mid \operatorname{card}(o_G(x)) \leq n\} \quad .$$

The latter set is clearly a closed subset of X. For a subgroup K of

G we use the notations, $\mathrm{Irr}(K)$ as the set of all irreducible unitary representations of K and $\mathrm{Irr}_n(K)$ as that of n-dimensional irreducible unitary representations of K. We denote each representation simply as u or more precisely as the map: $s \to u_s$. Let u be a unitary representation of K on the space H_u. Write the left coset space $G/K = \{s_\alpha K\}$ for the representatives $S = \{s_\alpha\}$ where $s_0 = e$. Let H_0 be a Hilbert space such that $\dim H_0 = |G/K|$. Put $H = H_0 \otimes H_u$, then each vector ξ in H is expanded as $\sum_\alpha e_\alpha \otimes \xi_\alpha$ with respect to a fixed complete orthonormal basis $\{e_\alpha\}$ in H_0 where the sum is actually ranging over countable numbers from indexes α for which $\xi_\alpha \neq 0$ together with the condition $\sum_\alpha \|\xi_\alpha\|^2 < \infty$. Define the unitary representation L_u^S of G on H induced by u in the following way:

$$L_u^S(s)(e_\alpha \otimes \xi) = e_\beta \otimes u_t \xi$$

if $ss_\alpha = s_\beta t$ for $t \in K$. One may then verify that this is actually a unitary representation of G. Now we take a point x in X with isotropy subgroup G_x. Taking the above subgroup K as G_x we may write

$$o_G(x) = \{s_\alpha x\}_{\alpha \in I} \quad .$$

Let π_x^S be the representation of $C(X)$ on H defined as

$$\pi_x^S(f)(e_\alpha \otimes \xi) = f(s_\alpha x) e_\alpha \otimes \xi \quad , \quad f \in C(X) \quad .$$

Let L_u^S be the representation of G defined as above for a unitary representation u of G_x on H_u. One then sees that (π_x^S, L_u^S) is a covariant representation of $\{C(X), G, \alpha\}$ and we can consider the representation of A_Σ defined by $\pi_{x,u}^S = \pi_x^S \times L_u^S$. As we have seen in Sec. 3.3 the GNS-representation π_φ of a pure state extension φ of μ_x has this structure once we notice the identification of the relevant Hilbert spaces,

$$H_\varphi = \sum_{s_\alpha} \otimes H_{s_\alpha} = H_0 \otimes H_u$$

in which the subspace H_e of H corresponds to the subspace $e_0 \otimes H_u$ and H_{s_α} to $e_\alpha \otimes H_u$ with $sx = s_\alpha x$.

<u>Lemma 4.1.1.</u> Keep the above notations. Then the representation $\pi^S_{x,u}$ does not depend on the choice of the representatives $S = \{s_\alpha\}$ of the left coset space G/G_x within unitary equivalence.

<u>Proof.</u> Let $R = \{r_\alpha\}$ be another representative for G/G_x and put $s_\alpha = r_\alpha t_\alpha$ for $t_\alpha \in G_x$. Let $\pi^S_{x,u}$ and $\pi^R_{x,u}$ be the corresponding representations of A_Σ on the space $H = H_0 \otimes H_u$. Define the unitary operator w on H by

$$w(e_\alpha \otimes \xi) = e_\alpha \otimes u_{t_\alpha}\xi \quad .$$

We assert that w intertwines the covariant representations (π^S_x, L^S_u) and (π^R_x, L^R_u). Indeed, as $s_\alpha x = r_\alpha x$ for every α one sees first that

$$w\pi^S_x(f)(e_\alpha \otimes \xi) = \pi^R_x(f)w(e_\alpha \otimes \xi) \quad .$$

On the other hand, take an element $s \in G$ and put $ss_\alpha = s_\beta t$ then

$$wL^S_u(s)(e_\alpha \otimes \xi) = w(e_\beta \otimes u_t\xi) = e_\beta \otimes u_{t_\beta t}\xi \quad ,$$

whereas

$$L^R_u(s)w(e_\alpha \otimes \xi) = L^R_u(s)(e_\alpha \otimes u_{t_\alpha}\xi)$$

$$= e_\beta \otimes u_{t_\beta t t_\alpha^{-1}} \cdot u_{t_\alpha}\xi$$

$$= e_\beta \otimes u_{t_\beta t}\xi$$

because $sr_\alpha = r_\beta t_\beta t t_\alpha^{-1}$. This completes the proof.

Definition 4.1.1. We write $\pi_{x,u}$ meaning any of the representation $\pi^S_{x,u}$ and call it, as well as, the representation (π_x, L_u) the representation of A_Σ or the covariant representation of $\{C(X), G, \alpha\}$ induced by the isotropy subgroup G_x. When the orbit of x is a finite set we say that the representation $\pi_{x,u}$ is of finite type.

Proposition 4.1.2. The representation $\pi_{x,u}$ is irreducible if and only if the representation u of G_x is irreducible.

Proof. Suppose a bounded operator b on H commutes with every element of $\pi^S_{x,u}(A_\Sigma)$, that is, $b \in \pi^S_{x,u}(A_\Sigma)'$. Put $b(e_\alpha \otimes \xi) = \Sigma e_\alpha \otimes \eta_\alpha$. If $\eta_\beta \neq 0$ with $\beta \neq 0$, we can choose a function $f \in C(X)$ such that $f(x) = 1$ and $f(s_\beta x) = 0$. This however contradicts the relation,

$$b(e_0 \otimes \xi) = b\pi^S_x(f)(e_0 \otimes \xi)$$

$$= \pi^S_x(f)(\sum_\alpha e_\alpha \otimes \eta_\alpha) = \sum_{\alpha \neq \beta} f(s_\alpha x)e_\alpha \otimes \eta_\alpha \quad .$$

This shows that b leaves the subspace $e_0 \otimes H_u$ invariant. Thus, it induces an operator $\bar{b}$ on H_u. Moreover, since b commutes with $L^S_u(t)$ for $t \in G_x$ the operator $\bar{b}$ commutes with u_t on H_u, hence $\bar{b} = \lambda 1_u$. Therefore, for every index α and $\xi \in H_u$ we have

$$b(e_\alpha \otimes \xi) = bL^S_u(s_\alpha)(e_0 \otimes \xi) = L^S_u(s_\alpha)b(e_0 \otimes \xi)$$

$$= \lambda L^S_u(s_\alpha)(e_0 \otimes \xi) = \lambda(e_\alpha \otimes \xi) \quad ,$$

that is, $b = \lambda 1$. Namely, $\pi_{x,u}$ is irreducible.

The converse is rather trivial.

The unitary equivalence of this kind of representation is given by the following

Theorem 4.1.3. Take two points x and y in X, then the representations $\pi_{x,u}$ and $\pi_{y,v}$ are unitarily equivalent if and only if $o(x) = o(y)$ and the representations of G_x, $t \to u_t$ and

$t \to v_{s_{\alpha_0}^{-1} t s_{\alpha_0}}$ are unitarily equivalent where $x = s_{\alpha_0} y$.

Proof. Suppose that $\pi^R_{x,u}$ and $\pi^S_{y,v}$ are equivalent by the unitary operator w for $R = \{r_\alpha\}$ and $S = \{s_\beta\}$. Let $K_1 = H_1 \otimes H_u$ and $K_2 = H_2 \otimes H_v$ be their acting spaces with the fixed basis $\{e_\alpha\}$ and $\{f_\beta\}$ in H_1 and H_2 respectively. Write

$$w(e_0 \otimes \xi) = \Sigma f_\beta \otimes \eta_\beta \quad .$$

If $o(x) \neq o(y)$, then $x \notin o(y)$ so that, for any index γ, with $\eta_\gamma \neq 0$, we can choose a function f such that $f(x) = 1$ and $f(s_\gamma y) = 0$. We then have

$$\Sigma f_\beta \otimes \eta_\beta = w(e_0 \otimes \xi) = w\pi^R_x(f)(e_0 \otimes \xi)$$

$$= \pi^S_y(f)w(e_0 \otimes \xi) = \sum_{\beta \neq \gamma} f(s_\beta y) f_\beta \otimes \eta_\beta$$

a contradiction. Hence $o(x) = o(y)$ with $x = s_{\alpha_0} y$. We shall show that the representations of G_x, u_t and $v_{s_{\alpha_0}^{-1} t s_{\alpha_0}}$ are equivalent.

By the above argument we have seen that w maps $e_0 \otimes H_u$ to $f_{\alpha_0} \otimes H_v$ and this defines a unitary operator w_1 of H_u to H_v. Then, for an element $t \in G_x$,

$$wL^R_u(t)(e_0 \otimes \xi) = w(e_0 \otimes u_t \xi) = f_{\alpha_0} \otimes w_1 u_t \xi$$

$$= L^S_v(t)w(e_0 \bar{\otimes} \xi) = L^S_v(t)(f_{\alpha_0} \otimes w_1 \xi)$$

$$= f_{\alpha_0} \bar{\otimes} v_{s_{\alpha_0}^{-1} t s_{\alpha_0}} w_1 \xi$$

because $ts_{\alpha_0} = s_{\alpha_0} \cdot s_{\alpha_0}^{-1} t s_{\alpha_0}$ and $s_{\alpha_0}^{-1} t s_{\alpha_0} \in G_y$. Therefore,

$$v_{s_{\alpha_0}^{-1}ts_{\alpha_0}} w_1 = w_1 u_t \quad .$$

Namely, w_1 intertwines two representations of G_x.

Suppose next that $O(x) = O(y)$ and the representations u_t and $v_{s_{\alpha_0}^{-1}ts_{\alpha_0}}$ for $x = s_{\alpha_0} y$ are unitarily equivalent by the unitary operator w_1. By (4.1.1) we can specify the representatives of the space G/G_x as $R = \{s_{\alpha_0} s_\alpha s_{\alpha_0}^{-1}\}$ together with the representatives $S = \{s_\alpha\}$ for G/G_y and moreover we may assume that $H_1 = H_2 \equiv H_0$ with common canonical base $\{e_\alpha\}$. Define the unitary operator w of K_1 to K_2 by

$$w(e_\alpha \otimes \xi) = L_v^S(s_{\alpha_0})(e_\alpha \otimes w_1\xi) \quad \text{for each} \quad e_\alpha \quad .$$

Put $s_{\alpha_0} s_\alpha = s_\beta t_0$ for $t_0 \in G_y$. We shall show that w intertwines (π_x^R, L_u^R) and (π_y^S, L_v^S). Indeed,

$$\begin{aligned} \pi_y^S(f)w(e_\alpha \otimes \xi) &= \pi_y^S(f)(e_\beta \otimes v_{t_0} w_1 \xi) \\ &= f(s_\beta y)(e_\beta \otimes v_{t_0} w_1 \xi) \\ &= w(f(s_\beta y)e_\alpha \otimes \xi) \\ &= w(\pi_x^R(f)(e_\alpha \otimes \xi)) \quad . \end{aligned}$$

Moreover, if $ss_{\alpha_0} s_\alpha s_{\alpha_0}^{-1} = s_{\alpha_0} s_\beta s_{\alpha_0}^{-1} t$ for $t \in G_x$,

$$wL_u^R(s)(e_\alpha \otimes \xi) = w(e_\beta \otimes u_t \xi) = L_v^S(s_{\alpha_0})(e_\beta \otimes w_1 u_t \xi)$$

and since $s_{\alpha_0}^{-1} ss_{\alpha_0} s_\alpha = s_\beta s_{\alpha_0}^{-1} ts_{\alpha_0}$

$$L_v^S(s)w(e_\alpha \otimes \xi) = L_v^S(s)L_v^S(s_{\alpha_0})(e_\alpha \otimes w_1\xi)$$

$$= L_v^S(s_{\alpha_0})L_v^S(s_{\alpha_0}^{-1}ss_{\alpha_0})(e_\alpha \otimes w_1\xi)$$

$$= L_v^S(s_{\alpha_0})(e_\beta \otimes v_{s_{\alpha_0}^{-1}ts_{\alpha_0}} w_1\xi)$$

$$= L_v^S(s_{\alpha_0})(e_\beta \otimes w_1u_t\xi) \quad .$$

This completes the proof.

Now suppose that G is abelian, then the representation $t \to v_{s_{\alpha_0}^{-1}ts_{\alpha_0}}$ reduces simply to the representation v_t with $G_x = G_y$. Moreover when u and v are irreducible they are nothing but characters χ_u and χ_v of G_x. Hence we have the following

<u>Corollary 4.1.4.</u> Assume that G is abelian. Then $\pi_{x,u}$ and $\pi_{y,v}$ are unitarily equivalent if and only if $O(x) = O(y)$ and the representations u and v of G_x are unitarily equivalent. If u and v are irreducible, this means that $\chi_u = \chi_v$.

Let (π, u) be a covariant representation of a dynamical system $\{C(X), G, \alpha\}$ on H. Write $\mathrm{ad}u_s(a) = u_s a u_s^*$ for $s \in G$. Let X_π be the spectrum of the C*-algebra $\pi(C(X))$. Denote the kernel of π by I_π, which is an α-invariant ideal of $C(X)$. There is then an invariant closed subset X_π' of X such that $I_\pi = k(X_\pi')$ where $k(X_\pi')$ means the set of all functions of $C(X)$ vanishing on X_π'. Moreover the quotient algebra $C(X)/I_\pi$ is naturally identified with $C(X_\pi')$. It follows that we may identify $\pi(C(X))$ with $C(X_\pi')$ as well as X_π with X_π'. With these identifications we may write that

$$\pi(f)(x) = f(x) \quad \text{on} \quad X_\pi \quad , \quad \|\pi(f)\| = \|f|X_\pi\| \quad .$$

Note that in these ways the restriction of the action of G to X_π' coincides with that of G on X_π induced by the group of automorphisms

$\{ad u_s | s \in G\}$ on $\pi(C(X))$. Henceforth we keep on this observation.

Proposition 4.1.5. Let $\tilde{\pi} = \pi \times u$ be an irreducible representation of A_Σ and let X_π be the spectrum of $\pi(C(X))$. Then for two nonempty open sets U and V in X_π there exists an element s in G such that $sU \cap V \neq \phi$.

Thus, when $G = Z$ the dynamical system $(X_\pi, \sigma|X_\pi)$ is topologically transitive.

Proof. Suppose that there exist nonempty open sets U and V with $sU \cap V = \phi$ for every $s \in G$. Then there exists a nonempty closed invariant set S with $S \cap V = \phi$. This yields a proper invariant ideal $I = k(S)$ of $\pi(C(X))$. It follows that $|IH| = H$ because $\tilde{\pi}$ is an irreducible representation. But this is a contradiction.

Definition 4.1.2. We call a representation (π, H) of A_Σ discrete if there exists a common eigenvector in H for all $\pi(f)(f \in C(X))$.

Every induced irreducible representation $\pi_{x,u}$ is discrete and actually this class of representations exhausts the class of discrete irreducible representations. Namely we have

Proposition 4.1.6. An irreducible representation (π, H) of A_Σ is discrete if and only if π is unitarily equivalent to an irreducible representation of the form $\pi_{x,u}$ for a point $x \in X$ and an irreducible unitary representation u of the isotropy subgroup G_x.

Proof. Let ξ_0 be a common unit eigenvector for $\pi(f)(f \in C(X))$ and define the pure state φ of A_Σ by

$$\varphi(a) = (\pi(a)\xi_0, \xi_0) \odot \quad .$$

By the property of ξ_0 the restriction of φ to $C(X)$ turns out to be a character of $C(X)$. Hence there exists a point x such that $\varphi(f) = f(x)$ for every $f \in C(X)$. Note that ξ_0 is a cyclic vector for $\pi(A_\Sigma)$ because π is irreducible. Therefore the representation π is unitarily equivalent to the GNS-representation $\{\pi_\varphi, H_\varphi, \xi_\varphi\}$ of φ by the unitary operator w defined as

$$w : \pi(a)\xi_0 \to \pi_\varphi(a)\xi_\varphi \qquad a \in A_\Sigma \quad .$$

Here the latter representation is shown to be of the form $\pi_{x,u}$ in Sec. 3.3.

If π is equivalent to some representation $\pi_{x,u}$ of finite type we also say that π is of finite type. We remark that in general a discrete representation has many common eigenvectors from points of different orbits.

An irreducible representation of A_Σ is not necessarily discrete in general. For instance, the representation of the algebra A_Σ which will be appearing in the proof of (4.3.4) is apparently not discrete but irreducible. We however have the following

Proposition 4.1.7. Every finite dimensional irreducible representation of A_Σ is discrete and of finite type.

The proof of this proposition is essentially included in the proof of the next theorem.

Theorem 4.1.8. Every irreducible representation of A is discrete and of finite type if and only if every orbit of G in X is a finite set.

The result states that if all orbits in X are finite sets, the Effros-Hahn conjecture is true in a very strong sense without any countability assumptions for groups and acting spaces.

Proof. Suppose that $X = \bigcup_{n=1}^{\infty} X_n$ and let π be an irreducible representation of A_Σ. From the assumption, $X_\pi = \bigcup_{n=1}^{\infty} (X_\pi)^n$ hence by the category theorem there exists a set, say $(X_\pi)^n$, having nonempty interior. Assume that n should be the smallest integer for which $(X_\pi)^n$ has such a property. Then the set $(X_\pi)_n = (X_\pi)^n \sim (X_\pi)^{n-1}$ has an interior point x. Write

$$o(x) = \{x, s_1x, s_2x, \ldots, s_{n-1}x\} \quad .$$

Choose a compact neighborhood U of x contained in $(X_\pi)_n$ such that the sets $\{s_iU\}$ are mutually disjoint. Since orbit of every point of U consists of n points, we see that the union of all s_iU's is an invariant closed set in X_π, which coincides with X_π by (4.1.5). We assert that U consists of just one point. In fact, if the interior of U contains other point y we can choose another compact neighborhood V in U which does not contain y and the above argument shows that $X = \bigcup_{i=0}^{n-1} s_iV$, a contradiction. Thus, $X_\pi = O(x)$ and each s_ix is an isolated point of X_π. Let p_i be the characteristic function of $\{s_ix\}$, which is a minimal projection in $\pi(C(X))$ and put $p_0 = \pi(f)$. One then verifies easily that $\pi(\alpha_{s_i}(f)) = p_i$ and the unitary representation u of G_x arising from π leaves p_0H invariant. Hence, u must be irreducible on p_0H and moreover $\pi(\delta_{s_i})p_0H = p_iH$. Thus π is unitarily equivalent to the representation $\pi_{x,u}$ on the space $H_n \otimes p_0H$ where H_n is an n-dimensional Hilbert space.

The converse is almost clear from (4.1.6) and (4.1.3) once we assume that every irreducible representation is discrete and of finite type.

Definition 4.1.3. A locally compact group G is called a Moore group if all irreducible unitary representations of G are finite dimensional.

When G is discrete the following four conditions are known to be equivalent;

(1) G is a Moore group;
(2) All irreducible representations of G are finite dimensional with bounded degree;
(3) G is of type I;
(4) G possesses an abelian subgroup of finite index.

A subgroup of a Moore group is a Moore group, too. With this concept for G we obtain

Corollary 4.1.9. Suppose G is a Moore group. Then every irreducible representation of A_Σ is finite dimensional if and only if

$X = \bigcup_{n=1}^{\infty} X_n$.

In this case A_Σ becomes a liminal C*-algebra.

Example 4.1.1. Let $X = \pi_{i \in Z}\{0, 1\}$ and let G be the group consisting of elements s in the restricted product of Z with pointwise addition mod 2 and with the property $\sum_n s_n = 0$. The group G is clearly a Moore group. Define the action of G on X by

$$sx = s(x_n) = (x_n + \langle x, s\rangle) \quad ,$$

where $\langle x, s\rangle = \Sigma x_n s_n$. The map s is injective because if $sx = sy$ we have

$$\langle x, s\rangle = \langle sx, s\rangle = \langle sy, s\rangle = \langle y, s\rangle$$

and $x_n = y_n$ for every n. Thus s is clearly a homeomorphism of X. Besides, since $\langle x, s\rangle = 1$ or 0, the orbit of a point $x = (x_n)$ of X consists of two points $\{(x_n), (x_n + 1)\}$ except two fixed points $x_0 = 0$ and $x_1 = (1)$.

As a special case we see that when $G = Z$ every irreducible representation of A_Σ is finite dimensional if and only if every point of X is a periodic point for the map σ.

We shall study next the case where the set $\bigcup_{n=1}^{\infty} X_n$ is dense in X. We shall treat such a case in Sec. 4.4.

Theorem 4.1.10. If the C*-algebra A_Σ has sufficiently many discrete irreducible representations of finite type, then the set $\bigcup_{n=1}^{\infty} X_n$ is dense in X. The converse holds if G is amenable.

Proof. Suppose that $\bigcup_{n=1}^{\infty} X_n$ is not dense in X. There exists then a nonzero function f in $C(X)$ vanishing on $\bigcup_{n=1}^{\infty} X_n$. By the definition with (4.1.6) every discrete irreducible representation of A_Σ of finite type kills f. Therefore the first assertion of the theorem holds.

Next suppose that G is amenable and $\bigcup_{n=1}^{\infty} X_n$ is dense in X. Let E be the canonical projection of norm one of A_Σ to $C(X)$, which is faithful in this case. For a point x in $\bigcup_{n=1}^{\infty} X_n$ let I_x be the ideal of A_Σ which is the intersection of all kernels of discrete irreducible representations of A_Σ associated with x. Let J be the intersection of all kernels of discrete irreducible representation of A_Σ of finite type. By (4.1.6) J is the intersection of all I_x's. Now suppose that J contains a nonzero element a. Since $a^*a \neq 0$ in J we may assume that a is positive and $E(a) \neq 0$. By the property of E in Sec. 3.2, the closure $\overline{E(I_x)}$ is a nonzero closed ideal of $C(X)$. Hence we can write as $\overline{E(I_x)} = k(S)$ for an invariant closed set S in X. However, since $E(I_x)$ contains the ideal $k(O(x))$, $S \subseteq O(x)$ and $S = O(x)$ because S is invariant. It follows that every function in $E(I_x)$ vanishes on $O(x)$. Therefore, $E(a)$ vanishes on every finite orbit so that $E(a) = 0$. This is a contradiction and the proof is completed.

Since a discrete Moore group is amenable, we obtain the next corollary corresponding to (4.1.9).

<u>Corollary 4.1.11.</u> Suppose that G is a Moore group. Then A_Σ has sufficiently many finite dimensional irreducible representations if and only if the set $\bigcup_{n=1}^{\infty} X_n$ is dense in X.

Let X/G be the orbit space with the quotient topology. We set $\tilde{X}/G$ the space of the equivalence relation in X/G with the quotient topology where two points p and q in X/G are equivalent if they have the same closure. Then we see that $\tilde{X}/G$ is a <u>compact</u> T_0-space. Each point in $\tilde{X}/G$ may be identified with the set $\overline{O(x)}$ for some point $x \in X$. We recall here that a primitive ideal of a C*-algebra is the kernel of an irreducible representation. Let $\mathrm{Prim}(A_\Sigma)$ be the set of all primitive ideals of A_Σ with hull kernel topology. Take a primitive ideal P of A, then it determines an invariant closed set F_P in X such that $P \cap C(X) = k(F_P)$. Namely, F_P is the hull of the ideal $P \cap C(X)$.

Lemma 4.1.12. Suppose that X is metrizable and G is countable. Then for each ideal P in $\mathrm{Prim}(A_\Sigma)$ there exists a point $x \in F_P$ such that $F_P = \overline{O(x)}$.

Proof. Let $\tilde{\pi} = \pi \times u$ be an irreducible representation of A_Σ such that $P = \tilde{\pi}^{-1}(0)$. We have then $F_P = X_{\tilde{\pi}}$. Since G is countable we can apply here the same argument as in the implication (3)$\Rightarrow$(4) in (1.1.3) and obtain the conclusion.

With the above assumption for X and G we can define a map π from $\mathrm{Prim}(A_\Sigma)$ to $\tilde{X}/G$ by $\pi(P) = F_P = \overline{O(x)}$.

Proposition 4.1.13. The map π is a continuous onto map.

Proof. Take a closed subset F in $\tilde{X}/G$. Then the inverse image F_1 in X is closed and invariant. Let J be the closed ideal of A_Σ generated by the ideal $k(F_1)$ of $C(X)$. Let $h(J)$ be the hull of J in $\mathrm{Prim}(A_\Sigma)$. We assert that $h(J) = \pi^{-1}(F)$, which shows the continuity of π. Thus let $\tilde{\pi} = \pi \times u$ be an irreducible representation such that $P = \tilde{\pi}^{-1}(0) \in \pi^{-1}(F)$. Then

$$k(F_1) \subseteq \pi^{-1}(0) = P \cap C(X) \quad .$$

Hence, $J \subseteq P$ and $P \in h(J)$. Conversely if $P \in h(J)$,

$$k(F_1) \subseteq J \cap C(X) \subseteq P \cap C(X) = k(F_P) \quad ,$$

which implies that $F_P \subseteq F_1$. Therefore, P belongs to $\pi^{-1}(F)$ and $h(J) = \pi^{-1}(F)$. On the other hand, for each set $\overline{O(x)}$ if we consider a discrete irreducible representation $\tilde{\pi}$ of A_Σ associated with a pure state extension of the character μ_x we have that $\pi(\tilde{\pi}^{-1}(0)) = \overline{O(x)}$. This completes the proof.

For each point x the constant function 1 on the isotropy subgroup G_x gives rise to a discrete irreducible representation $\pi_{x,u}$ by (3.3.7). Since, in this case, the unitary representation u of G_x is just the trivial representation, the representation $\pi_{x,u}$ depends only on the orbit of x by (4.1.3). Now define the map π_0:

$X \to \mathrm{Prim}(A_\Sigma)$ by $\pi_0(x) = \pi_{x,u}^{-1}(0)$. In some cases such as the case where X is metrizable and all isotropy subgroups are abelian, it can be proved that the map π_0 is continuous. In such a case the restriction of π to $\pi_0(X)$ induces a one-to-one map onto $\tilde{X}/G$.

4.2 Topological Equivalence Between Finite Orbit Structure and Spaces of Finite Dimensional Irreducible Representations of Transformation Group C*-algebras

In this section simplifying our discussion to the case $G = Z$, $\Sigma = (X, \sigma)$ we shall show the topological equivalence between finite orbit structure for σ and spaces of finite dimensional irreducible representations of $A_\Sigma = C(X) \rtimes_{\alpha\ r} Z$. Let H_n be a fixed n-dimensional Hilbert space with an orthogonal basis $\{e_0, e_1, \ldots, e_{n-1}\}$. We denote $\mathrm{Irr}_n(A_\Sigma)$ the set of all irreducible representations of A on H_n endowed with the pointwise strong convergence topology. Let $\hat{A}_{\Sigma n}$ be the set of unitary equivalence class of n-dimensional irreducible representations of A_Σ. The space $\hat{A}_{\Sigma n}$ is considered as the quotient space of $\mathrm{Irr}_n(A_\Sigma)$ so that it becomes a topological space by the quotient topology. Incidentally, the set of all irreducible representations of a C*-algebra A is usually denoted by $\mathrm{Irr}(A)$ and its unitary equivalence class by $\hat{A}$. The latter is called the dual of A. As in $\hat{A}_{\Sigma n}$ there is a natural topology in A but the space becomes hardly a Hausdorff space.

Now by (4.1.7) an n-dimensional irreducible representation π of A_Σ is equivalent to the representation $\pi_{x,u}$ for a point $x \in X_n$ and an irreducible representation u of the isotropy subgroup G_x. However, since $G = Z$ in the present situation the representation u simply means a character of the group $G_x = \{kn \mid k \in Z\}$, which is identified with a complex number z in the unit circle S^1. We may also assume that $\pi_{x,u}$ is acting on the space H_n. Thus we write $\pi_{x,z}$ instead of $\pi_{x,u}$ or (π_x, u_z) as a covariant irreducible representation of $\{C(X), Z, \alpha\}$ on H_n, where the unitary u_z is specified as

$$u_z e_k = e_{k+1}\ (0 \le k \le n-2) \quad \text{and} \quad u_z e_{n-1} = z e_0 \quad .$$

Let X_n/Z be the orbit space for X_n. From (4.1.3) and (4.1.7) we confirm the following facts. Namely,

$$\mathrm{Irr}_n(A_\Sigma) = \{\mathrm{ad}u(\pi_{x,z}) \mid (x, z) \in X_n \times S^1,\ u \in U(H_n)\}$$

and the map

$$\Phi: (x, z) \in X_n \times S^1 \to \pi_{x,z} \in \mathrm{Irr}_n(A_\Sigma)$$

induces a one-to-one and onto map

$$\Psi: (X_n/Z) \times S^1 \to \hat{A}_{\Sigma n} \quad .$$

We note that the algebra $\pi_x(C(X))$ forms the diagonal algebra with respect to the basis $\{e_i\}$, hence it is a maximal abelian C*-subalgebra of $B(H_n)$.

Theorem 4.2.1. The map Ψ is a homeomorphism between the spaces $(X_n/Z) \times S^1$ and $\hat{A}_{\Sigma n}$.

Proof. Let (x, z) and (y, w) be in $X_n \times S^1$. To show that Ψ is continuous it suffices to see that the map Φ is continuous. Take a function f of $C(X)$, then for $0 \le k \le n-2$

$$\| \pi_x(f)u_z^{in+k}e_0 - \pi_y(f)u_w^{in+k}e_0\|$$

$$= \| f(\sigma^k(x))z^i e_k - f(\sigma^k(y))w^i e_k\|$$

$$= |f(\sigma^k(x))z^i - f(\sigma^k(y))w^i| \to 0$$

$$\text{as} \quad x \to y \quad \text{and} \quad z \to w \quad .$$

We get similar conclusion on the other vectors $\{e_k\}$. It follows that for every element a in $k(Z, C(X))$ and $\xi \in H_n$,

$$\| \pi_{x,z}(a)\xi - \pi_{y,w}(a)\xi\| \to 0 \quad \text{as} \quad x \to y \quad \text{and} \quad z \to w \quad .$$

Since $k(Z, C(X))$ is dense in A_Σ we obtain the same strong convergence for every element of A_Σ. Thus Φ is continuous. We shall show next that Ψ is an open map.

$$\begin{array}{ccc} X_n \times S^1 & \xrightarrow{\Phi} & \mathrm{Irr}_n(A) \\ \downarrow & & \downarrow q \\ (X_n/Z) \times S^1 & \xrightarrow{\Psi} & \hat{A}_{\Sigma n} \end{array}$$

Let q be the quotient map from $\mathrm{Irr}_n(A_\Sigma)$ to $\hat{A}_{\Sigma n}$. Then it suffices to prove that the map $q \cdot \Phi$ is an open map. Thus, let U be an open set in $X_n \times S^1$. In order to show that $q \cdot \Phi(U)$ is open in $\hat{A}_{\Sigma n}$ we shall show that the set $W = q^{-1} \cdot q \cdot \Phi(U)$ is open in $\mathrm{Irr}_n(A_\Sigma)$. Note first that by definition of q

$$W = \{ad u(\pi_{x,z}) \mid (x, z) \in U,\ u \in U(H_n)\} \quad .$$

An element $\rho = ad u \cdot \pi_{x,z}$ is an interior point of W if and only if $\pi_{x,z}$ is an interior point of W. Hence we assert that every element $\Phi(x, z) = \pi_{x,z}$ for $(x, z) \in U$ is an interior point of W. Choose $\varepsilon > 0$ and a neighborhood $U(x)$ of x in X such that

$$U^0 = \{(y, w) \in X_n \times S^1 \mid y \in U(x),\ |z - w| < \varepsilon\} \subset U \quad .$$

Let f be the function of $C(X)'$ such that $f(x) = 1$ and $f(y) = 0$ for $y \notin U(x)$. Take two elements of A_Σ, $\hat{f} = f\delta_0$ and δ_n, and let W^0 be a neighborhood of $\Phi(x, z)$ every element ρ of which satisfies the following conditions;

$$\| \rho(\hat{f})e_0 - \pi_{x,z}(\hat{f})e_0 \| < 1 \quad ,$$

$$\| \rho(\delta_n)e_0 - \pi_{x,z}(\delta_n)e_0 \| < \varepsilon \quad .$$

Write $\rho = ad u \cdot \pi_{y,w}$, then

$$\| u\pi_y(f)u^*e_0 - e_0 \| = \| \rho(\hat{f})e_0 - e_0 \| < 1 \quad .$$

Hence, $\pi_y(f) \neq 0$, that is,

$$\pi_y(f)e_k = f(\sigma^k(y))e_k \neq 0 \quad \text{for some } k \quad .$$

This means that $\sigma^k(y) \in U(x)$. On the other hand, the second condition for ρ says that

$$\| u u_w^n u^* e_0 - u_z^n e_0 \|$$

$$= \| we_0 - ze_0 \| = |w - z| < \varepsilon \quad .$$

Therefore, $(\sigma^k(y), w) \in U^0$ and $q(\rho) \in q \cdot \Phi(U^0)$ because $\pi_y \times u_w$ and $\pi_{\gamma k(y)} \times u_w$ are unitarily equivalent. Thus,

$$W^0 \subset q^{-1} \cdot q \cdot \Phi(U^0) \subset W \quad .$$

This completes the proof.

Let $X(k) = \pi_{i \in Z}\{0,1,2,\ldots,k-1\}$ be the symbolic dynamical system of k-elements with Bernoulli shift σ_k. Then

$$X(k)_1 = \{x \in X(k) | \sigma_k(x) = x\}$$

$$= \{x = (x_i)_{i \in Z} | x_i = x_0 \text{ for all } i \in Z\} \quad .$$

We denote $A(k)$ the C*-algebra associated with this dynamical system. By the above theorem $A(k)_1$ is homeomorphic with the k-fold topological sum of the unit circle S^1, i.e. $A(k)_1 \cong S^1 \oplus S^1 \oplus \ldots \oplus S^1$. Therefore, these C*-algebras are classified by their indexes. Namely, we have

<u>Theorem 4.2.2.</u> The C*-algebra $A(k)$ is *-isomorphic to $A(j)$ if and only if $k = j$.

One may further formulate the above result into more general form. Indeed, let $X_K = \pi_{i \in Z}K$ for a compact space K and consider the shift $\sigma_K : (x_i)_{i \in Z} \to (x_{i-1})_{i \in Z}$ on X_K. Put $A_K = C(X_K) \rtimes_{\alpha r} Z$. Since $(X_K)_1$

is homeomorphic with K we see that:

"For two compact spaces K_1 and K_2, if A_{K_1} is $*$-isomorphic to A_{K_2} then $K_1 \times S^1$ and $K_2 \times S^1$ are homeomorphic to each other".

As in the case of the above theorem if those involved compact spaces are topologically nice spaces, we can derive further conclusion that K_1 is homeomorphic to K_2. But this is not the case in general. Indeed, it is known that there exist topological spaces K_1 and K_2 which are not homeomorphic but have homeomorphic product spaces $K_1 \times S^1$ and $K_2 \times S^1$.

4.3 Relations Between the Structure of Ideals of Transformation Group C*-algebras and the Orbit Structure of Their Dynamical Systems

In this section we shall discuss the structure of ideals in A_Σ for a dynamical system $\Sigma = (X, \sigma)$ in relation with the behavior of the homeomorphism σ. The argument involves the problems: under what condition for σ can it be concluded that $I \cap C(X) \neq \{0\}$ for every proper ideal I in A_Σ and in particular about the simplicity of the algebra A_Σ. In the following, we identify A_Σ with the reduced crossed product $C(X) \rtimes_{\alpha r} Z$ as in Sec. 4.2.

Let (π, u) be a covariant representation of $\{C(X), Z, \alpha\}$. Put $B = \pi \times u(A_\Sigma)$. Let X_π be the spectrum of $\pi(C(X))$.

<u>Theorem 4.3.1.</u> Keep the above notations. Assume that the dynamical system $(X_\pi, \sigma|X_\pi)$ is topologically transitive, then there exists a projection of norm one ε_π of B to $\pi(C(X))$ such that

$$\varepsilon_\pi \cdot \pi \times u(a) = \pi \cdot \varepsilon(a) \quad ,$$

provided that X_π consists of infinite points.

The representation $\pi \times u$ is faithful if and only if π is faithful on $C(X)$.

Proof. We show first the last version of the theorem. Suppose $\pi \times u(a) = 0$, then $\pi \times u(a^*a) = 0$. Hence, $\pi \cdot \varepsilon(a^*a) = 0$ and $\varepsilon(a^*a) = 0$ if π is faithful. It follows that $a^*a = 0$ and $a = 0$. As the converse is trivial we have the conclusion. To prove the theorem, it suffices to show the inequality

$$\| \pi(f_0) \| \leq \| \sum_{-n}^{n} \pi(f_k)u^k \|$$

for every element $a = \sum_{-n}^{n} \pi(f_k)u^k$ in B. Thus fix this element a and assume that $\pi(f_0) \neq 0$. Take an $\varepsilon > 0$ such that

$$\| \pi(f_0) \| - 2\varepsilon > 0 \quad .$$

We assert next that X_π contains an aperiodic point or periodic points of arbitrary high periods. In fact, if $X_\pi = \bigcup_{k=1}^{n} X_k$ some X_k must contain an open set U and since the system is topologically transitive the set $\bigcup_{i=0}^{k-1} \sigma^i(U)$ is dense in X_π. It follows that X_π consists of k-periodic points. Take $k+1$ points $\{x, \sigma x, \ldots, \sigma^{k-1}x, y\}$ and their disjoint neighborhoods $\{U_0, U_1, \ldots, U_{k-1}, V\}$. As in the proof of (4.1.8) we may assume that the union of U_i's is invariant but this is a contradiction.

Now choose disjoint open sets in X_π, $\{P_i\}$ $(-2n \leq i \leq 2n)$ such that $\sigma(P_j) = P_{j+1}$ for $-2n \leq j \leq 2n-1$. Let x_0 be a point of X with $|f_0(x_0)| = \| \pi(f_0) \|$ and take a neighborhood Q of x_0 such that

$$Q = \{x \in X_\pi \,|\, |f_0(x) - f_0(x_0)| < \varepsilon\} \quad .$$

Since P_0 and Q are nonempty open sets, there exists an integer m such that

$$\sigma^m(P_0) \cap Q \neq \phi \quad .$$

Replacing P_i by $\sigma^m(P_i)$ we may assume that $m=0$. Let g be a function of $C(X)$ satisfying the conditions:

$$\text{supp}\,(g|X_\pi) \subset P_0 \cap Q \quad , \qquad \|\pi(g)\| = \|g\,|X_\pi\| = 1$$

and

$$\|\pi(g)\pi(f_0)\| \geq \|\pi(f_0)\| - \varepsilon \quad .$$

Let H be the representation space of (π, u) and choose a unit vector ξ such that

$$\|\pi(gf_0)\xi\| \geq \|\pi(gf_0)\| - \varepsilon$$

$$> \|\pi(f_0)\| - 2\varepsilon > 0 \quad .$$

Note that here the vectors

$$\{\pi(f_k)u^k\pi(g)\xi \mid -n \leq k \leq n\}$$

are orthogonal. In fact, if $k \neq j$ for $-n \leq k, j \leq n$

$$(\pi(f_j)u^j\pi(g)\xi, \pi(f_k)u^k\pi(g)\xi)$$

$$= (\pi(\bar{g})u^{*k}\pi(\bar{f}_k f_j)u^j\pi(g)\xi, \xi)$$

$$= (u^{*k}\pi(\alpha^k(\bar{g})\bar{f}_k f_j)u^j\pi(g)\xi, \xi)$$

$$= (u^{*k}\pi(\bar{f}_k f_j)u^j\pi(\alpha^{k-j}(\bar{g})g)\xi, \xi) = 0 \quad .$$

Therefore,

$$\|a\pi(g)\xi\|^2 = \sum_{-n}^{n} \|\pi(f_k)u^k\pi(g)\xi\|^2$$

$$\geq \|\pi(f_0 g)\xi\|^2 \geq (\|\pi(f_0)\| - 2\varepsilon)^2 \quad .$$

Hence,

$$\|a\| \geq \|\pi(f_0)\| - 2\varepsilon \quad ,$$

which implies that $\|a\| \geq \|\pi(f_0)\|$.

With (4.1.5) we have an immediate following consequence.

Proposition 4.3.2. Let π be an infinite dimensional irreducible representation of A_Σ, then there exists a projection of norm one ε_π of $\pi(A_\Sigma)$ to $\pi(C(X))$ such that $\varepsilon_\pi \cdot \pi(a) = \pi \cdot \varepsilon(a)$.

We remark that in the case of a finite dimensional irreducible representation π, there also exists a projection of norm one from $\pi(A_\Sigma)$ to $\pi(C(X))$ because $\pi(C(X))$ coincides with the diagonal matrix algebra when $\pi(A_\Sigma)$ is regarded as the full matrix algebra. But this projection does not satisfy the above relation in general.

Another typical case to which the theorem applies is the case of a minimal dynamical system.

Theorem 4.3.3. The algebra A_Σ is simple if and only if $\Sigma = (X, \sigma)$ is minimal provided that X consists of infinite points.

Proof. Suppose that Σ is minimal and let (π, u) be a covariant representation of $\{C(X), Z, \alpha\}$. Since Σ is minimal, $X_\pi = X$ for this covariant representation and π is a *-isomorphism. Hence by the above theorem with (1.1.4) the representation $\pi \times u$ of A_Σ is a *-isomorphism. It follows that A_Σ is simple. For if there exists a proper ideal I in A it would give rise to a representation of A_Σ which is not faithful through the quotient algebra A_Σ/I.

Conversely suppose that A_Σ is simple and Σ is not minimal. We shall derive a contradiction. From the assumption there exists a proper σ-invariant closed subset in X, which determines a proper invariant ideal I_0 of $C(X)$. Let J be an ideal of A generated by I_0 and the unitaries δ_n. Now as I_0 is proper, by (2.1.1)

$$\| 1 - f \| \geq 1 \qquad \text{for every} \quad f \in I_0 \quad .$$

On the other hand, we have that $E(J) = I_0$, hence for every $a \in J$

$$1 \leq \| E(1 - a) \| \leq \| 1 - a \| \quad .$$

Therefore, J is a proper ideal, a contradiction.

The theorem says that when the system Σ is minimal we may choose any convenient representation of A_Σ to analyse its structure. A good candidate for this purpose is the covariant representation on the space $L^2(X, \mu)$ consisting of the representation of $C(X)$ as multiplication operators m_f (i.e. $m_f g = fg$ for $g \in L^2(X, \mu)$) and of the unitary representation of Z induced by σ.

Now consider the rigid irrational rotation σ_θ on the torus by an irrational number $\theta \in [0, 1)$ in Example 1.1.2. The algebra A_θ associated with the system (T, σ_θ) is at first separable and simple by the above theorem. Moreover it has unique tracial state τ as shown in Sec. 3.3. However, though a great deal of things is known about homeomorphisms on the torus T since Poincare's work but little was known about the structure of the transformation group C*-algebras associated with those homeomorphisms even about the irrational rotation C*-algebra A_θ until early 80's. Those C*-algebras such as A_θ occur in a variety of situations. For instance, we shall see later the irrational rotation algebras occur as the simple infinite dimensional quotients of the group C*-algebra of the 3-dimensional discrete Heisenberg group. It is known that they are exactly the simple C*-algebras on which the torus group T^2 has ergodic actions. Here we shall show another structure of A.

We should mention first that A_θ is exactly the C*-algebra generated by any pair of unitary operators u and v which satisfy the commutation relation $uv = e^{-2\pi i\theta}vu$. To see this, by the above theorem it is enough to show that such a C*-algebra is generated by a covariant representation of $\{C(T), \alpha_\theta\}$. Now since $uvu^* = \lambda v$ for $\lambda = e^{-2\pi i\theta}$,

$$sp(v) = sp(uvu^*) = \lambda sp(v) \quad ,$$

which means that, as a closed subset of S^1, $sp(v)$ contains an orbit of σ_θ. Therefore, $sp(v) = S^1$ and the commutative C*-algebra generated by v may be identified with $C(T)$ on which $adu(v) = \lambda v =$

$\alpha_\theta(v)$. Namely the C*-algebra $C^*(u, v)$ is generated by this covariant representation.

For some time it was believed that A_θ might not have projections. It has turned out however that A_θ is rich in projections.

Theorem 4.3.4. For any number $\alpha \in (Z + Z\theta) \cap [0,1]$, there exists a projection p in A_θ such that $\tau(p) = \alpha$.

Proof. We will view the elements of $C(T)$ as continuous functions on the real line R, which are periodic of period 1. Furthermore as remarked before we consider A_θ as the C*-algebra on $L^2(T)$ generated by multiplication operators m_f for the functions f of $C(T)$ and the shift unitary s defined on $L^2(T)$ by $s(f)(t) = f(t-\theta)$. Recall that on this algebra the value of the tracial state τ is given by

$$\tau\left(\sum_{-n}^{n} f_k s^k\right) = \tau(f_0) = \int_0^1 f(t)dt \quad .$$

We shall first construct a projection p with $\tau(p) = \theta$ in the form (supported on $[-1,0,1]$)

$$p = m_h s^{-1} + m_f + m_g s$$

for $h, f, g \in C(T)$. In the following we identify a function f with the multiplication operator m_f. Now suppose that we have a projection of the above form. From the condition $p = p^*$ one easily sees that f is real-valued and $h = s^*(g)$. With further idempotent condition we have finally the relations:

(1) $g(t)g(t-\theta) = 0$;

(2) $g(t)[1 - f(t) - f(t-\theta)] = 0$;

(3) $f(t)(1 - f(t)) = |g(t)|^2 + |g(t+\theta)|^2 \qquad t \in R.$

Conversely, if f and g are functions of $C(T)$ satisfying these conditions and if we put $h = s^*(\bar{g})$, we can verify that the corresponding element of A_θ is a projection. Henceforth we assume that

$\theta \in (0, \frac{1}{2})$ because s^* is regarded as the unitary induced by the translation $1-\theta$ and $A_\theta \cong A_{1-\theta}$. Choose any $\varepsilon > 0$ such that $\varepsilon < \theta$ and $\theta + \varepsilon < \frac{1}{2}$. Let f be a continuous function in $C(T)$ defined as follows. On the interval $[0, \varepsilon]$ f may be any continuous function with values in $[0, 1]$ and with $f(0) = 0$ and $f(\varepsilon) = 1$. The function f should be constant value 1 on $[\varepsilon, \theta]$ and 0 on $[\theta + \varepsilon, 1]$ respectively, and finally define f on $[\theta, \theta + \varepsilon]$ by

$$f(t) = 1 - f(t - \theta) \quad .$$

We define the function g by

$$g(t) = (f(t)(1 - f(t))^{\frac{1}{2}} \quad \text{on} \quad [\theta, \theta + \varepsilon]$$

and let g be zero elsewhere on $[0, 1]$. Then f and g satisfy relations (1), (2) and (3) above, so that they define a projection p in the above form. By the construction

$$\tau(p) = \int_0^1 f(t)dt = \theta \quad .$$

For the general case, we consider the following way. For a positive integer n, the algebra $C(T)$ contains the algebra $C_n(T)$ of continuous functions on R periodic of period $\frac{1}{n}$. On this algebra $C_n(T)$ the shift operator by θ behaves like the shift on $C(T)$ by $\{n\theta\}$, the fractional part of $n\theta$. It follows that $A_{\{n\theta\}}$ is embedded as a C*-subalgebra of A_θ. In this situation, the restriction of the tracial state to $A_{\{n\theta\}}$ will be the unique tracial state on $A_{\{n\theta\}}$ and so a projection in $A_{\{n\theta\}}$ of trace $\{n\theta\}$, constructed as above, will be a projection in A_θ of same trace (supported also on $[-1, 0, 1]$). The case of the fractional part $\{-n\theta\}$ will be finished by finding a projection of form $1 - \{-n\theta\} = \{n\theta\}$, handled as above. This completes the proof.

We shall see later (cf. (5.3.3)) that the range of τ is contained in $(Z+Z\theta)\cap[0, 1]$, so that it actually coincides with the set $(Z+Z\theta)\cap[0, 1]$. Moreover, we shall show that the group $Z+Z\theta$ plays a complete algebraic invariant for A_θ $(0<\theta<\frac{1}{2})$.

As mentioned before if an ideal I of A is induced from an ideal I_0 of $C(X)$, I is apparently an ideal with $I\cap C(X)\neq\{0\}$. But in general an arbitrary proper ideal of A_Σ does not necessarily have the nonzero intersection with $C(X)$. The Effros-Hahn conjecture introduced in 3.3 concerns this thing for primitive ideals.

Theorem 4.3.5. The following three conditions are equivalent;

(1) $C(X)$ is a maximal abelian C*-subalgebra of A_Σ;
(2) For each proper ideal I of A, $I\cap C(X)\neq\{0\}$;
(3) The interior of the set X^n is empty for all n.

Proof. The equivalence of (1) and (3) has been almost established in (3.3.3). We shall show the equivalence of (2) and (3). Suppose that the interior of X^{n0} is not empty for some integer n_0. Then the interior of $X_n\neq\phi$ for some integer n. Let U' be an open set contained in X_n and put

$$U = U'\cup\sigma U'\cup\ldots\ \sigma^{n-1}U'$$

which is an invariant open set still contained in X_n. Let f be a nonzero function of $C(X)$ supported in U. For a point x, let ρ_x be an irreducible representation of A_Σ on a space H_x associated with a pure state extension of the evaluation functional μ_x on $C(X)$. We recall the structure of (ρ_x, H_x) described in Sec. 3.3. With this structure we see that $\rho_x(f)=0$ for $x\in U^c$, whereas $\rho_x(f)\neq 0$ for some point x in U. Note that the family of irreducible representations $\{\rho_x | x\in X\}$ is faithful on $C(X)$. Thus in order to get the conclusion it is enough to find a nonzero element a in $\ell^1(Z, C(X))$ such that $\rho_x(a)=0$ for all x. Indeed, the ideal I generated by this element a has the property $I\cap C(X)=\{0\}$.

Now we specify ρ_x for a point x in U as the representation $\pi_{x,1}$ on the space H_n (cf. Sec. 4.2). Define the element a in

$\ell^1(Z, C(X))$ as

$$a = -f + f\delta_n \quad .$$

Then by definition of $\pi_{x,1}$ we have

$$\pi_{x,1}(a)\xi = -\pi(f)\xi + \pi(f)u_1^n\xi$$
$$= -\pi(f)\xi + \pi(f)\xi = 0 \qquad \xi \in H_n \quad .$$

Namely, $\pi_{x,1}(a) = 0$ for all x in U. Therefore, $\rho_x(a) = 0$ for all points x in X.

Next suppose that there exists a proper ideal I in A such that $I \cap C(X) = \{0\}$ under the assumption that the interior of X^n is empty for all n. Consider the quotient map q. The restriction of q to $C(X)$ is by assumption a *-isomorphism. For each point $x \in X$ we fix an irreducible representation ρ_x of A_Σ associated with a pure state extension $\varphi \cdot q$ where φ is a pure state extension on $B = A_\Sigma/I$ of the functional μ_x' on $q(C(X))$ defined by $\mu_x'(q(f)) = f(x)$. Take a nonzero element a of I and denote $\{a(n)\}$ the Fourier coefficients of a which are defined by $a(n) = \varepsilon(a\delta_n^*)$. As $a \neq 0$, there is an integer n_0 such that $a(n_0) \neq 0$ and $a(n_0)(x_0) = 0$ for some point x_0. Since I is an ideal, we may therefore assume that $a(0)(x_0) = 2$. Take a neighborhood U of x_0 such that $|a(0)(x)| \geq 1$ for $x \in U$. Here U contains either aperiodic points or periodic points with arbitrary high periods, for otherwise all points of U are periodic with bounded degree, say by m, and U is contained in the set $X^{m!}$, contradicting the assumption.

Now choose an element b of $\ell^1(Z, C(X))$ with $\| a - b \| < \frac{1}{4}$ as well as a number N_0 such that

$$\| b \|_1 \leq \sum_{-N_0}^{N_0} \| b(n) \| + \frac{1}{4} \quad .$$

Take a point x in U and assume first that x is a periodic point with period k. Then we may consider ρ_x as the representation $\pi_{x,z}$ acting on the space H_k whose associated pure state is given by $(\pi_{x,z}(a)e_0, e_0)$ (cf. Sec. 4.2). For each i with $1 \le i \le k-1$, we have

$$\sum_{n=-\infty}^{\infty} (\pi_{x,z}(b(i+nk))u_z^{i+nk}e_0, e_0)$$

$$= \sum_{-\infty}^{\infty} (b(i+nk)(\sigma^i(x))z^n e_i, e_0) = 0 \quad .$$

Therefore,

$$(\rho_x(b)e_0, e_0) = (\pi_{x,z}(b)e_0, e_0)$$

$$= (\pi(b(0))e_0, e_0) + \sum_{n\neq 0} z^n(b(nk)(x)e_0, e_0) \quad ,$$

whence

$$|b(0)(x)| = |(\pi(b(0))e_0, e_0)|$$

$$\le \| \rho_x(b) \| + \sum_{n\neq 0} \| \rho_x(b(nk)) \|$$

$$< \| \rho_x(b) \| + \frac{1}{4} \qquad \text{if} \quad k > N_0 \quad .$$

It follows that $|b(0)(x)| < \frac{1}{2}$ because

$$\| \rho_x(b) \| = \| \rho_x(b) - \rho_x(a) \|$$

$$\le \| b - a \| < \frac{1}{4} \quad .$$

When x is an aperiodic point, the isotropy subgroup is trivial and ρ_x is an induced representation of one dimensional (covariant) representation of $\{C(X), G_x\}$. Therefore, if e_0 is the starting base in the representation space of ρ_x we reach more easy calculation than before and get the identities;

$$(\rho_x(b)e_0, e_0) = \sum_{-\infty}^{\infty} (\rho_x(b(n))e_n, e_0)$$

$$= (\rho_x(b(0))e_0, e_0) = b(0)(x) \quad .$$

Hence along the same line as before we obtain $|b(0)(x)| < \frac{1}{2}$. Now since we can choose the neighborhood U arbitrarily small, this implies that $|b(0)(x_0)| \le \frac{1}{2}$. However it leads to the estimation,

$$|a(0)(x_0)| \le \frac{3}{4}$$

because

$$\| a(0) - b(0) \| \le \| a - b \| < \frac{1}{4} \quad .$$

This is a contradiction and we finish the proof.

It is to be noticed that we can also make use of this result to prove (4.3.3). Indeed, the one implication: minimality → simplicity, is rather natural in this case because with the condition of the theorem every *-homomorphism which is faithful on $C(X)$ becomes a *-isomorphism.

4.4 Representations of the Discrete 3-Dimensional Heisenberg Group. An Application

We shall apply our preceding arguments to the analysis of representations of the discrete 3-dimensional Heisenberg group H through its group C*-algebra. Write

$$H = \left\{ \begin{pmatrix} 1 & \ell & m \\ 0 & 1 & n \\ 0 & 0 & 1 \end{pmatrix} \middle| \ \ell, m, n \in Z \right\}$$

and recall that H is generated by two elements

$$a = \begin{pmatrix} 1 & 1 & 0 \\ 0 & 1 & 0 \\ 0 & 0 & 1 \end{pmatrix} \quad \text{and} \quad b = \begin{pmatrix} 1 & 0 & 0 \\ 0 & 1 & 1 \\ 0 & 1 & 1 \end{pmatrix} \quad .$$

Put

$$c = aba^{-1}b^{-1} = \begin{pmatrix} 1 & 0 & 1 \\ 0 & 1 & 0 \\ 0 & 0 & 1 \end{pmatrix} .$$

Then c commutes with a and b, thus it belongs to the center of H. Let G be the abelian subgroup generated by b and c, which is expressed as follows;

$$G = \left\{ \begin{pmatrix} 1 & 0 & \ell \\ 0 & 1 & n \\ 0 & 0 & 1 \end{pmatrix} \middle| \ \ell, n \in Z \right\} .$$

Since

$$\begin{pmatrix} 1 & 0 & \ell \\ 0 & 1 & n \\ 0 & 0 & 1 \end{pmatrix} \begin{pmatrix} 1 & 0 & \ell' \\ 0 & 1 & n' \\ 0 & 0 & 1 \end{pmatrix} = \begin{pmatrix} 1 & 0 & \ell + \ell' \\ 0 & 1 & n + n' \\ 0 & 0 & 1 \end{pmatrix} ,$$

the group G is isomorphic to the product group $Z \times Z$. On the other hand, we note that a is a normalizer of G. Let α be the automorphism of G defined by $\alpha(x) = axa^{-1}$, by which we can consider the action of Z on G. Now consider the decomposition of an element of H as

$$\begin{pmatrix} 1 & m & \ell \\ 0 & 1 & n \\ 0 & 0 & 1 \end{pmatrix} = \begin{pmatrix} 1 & 0 & \ell \\ 0 & 1 & n \\ 0 & 0 & 1 \end{pmatrix} \begin{pmatrix} 1 & m & 0 \\ 0 & 1 & 0 \\ 0 & 0 & 1 \end{pmatrix} ,$$

which we may write as (x, m) $(x \in G)$ because the correspondence

$$m \in Z \rightarrow \begin{pmatrix} 1 & m & 0 \\ 0 & 1 & 0 \\ 0 & 0 & 1 \end{pmatrix}$$

is a group isomorphism. One then easily verify that

$$(x, m)(x', m') = (x\alpha^m(x'), m + m') .$$

Thus $H = G \underset{\alpha}{\times} Z$, the semidirect product by the action of Z on G.

Hence, H is an amenable group. In order to analyse the group C*-algebra C*(H) we need the following

Proposition 4.4.1. Let $M = L \times_\alpha N$ be a discrete group which is a semidirect product of its subgroups L and N by the action of N on L as $\alpha_n(\ell) = n\ell n^{-1}$. Then the C*-algebra C*(M) is *-isomorphic to the crossed product $C^*(L) \rtimes_\alpha N$ where α is an action of N on C*(L) induced by that of N on L.

Proof. We shall define an isomorphism Φ between the algebra $\ell^1(M)$ (as a usual convolution algebra with *-operation $a^*(m) = a(m^{-1})$) and the algebra $\ell^1(N, \ell^1(L))$ (as a twisted convolution *-algebra for constructing the crossed product). For an element $a = (a(m))$ in $\ell^1(M)$ put $\Phi(a)(n) = a_n$ for $n \in N$ where a_n is a function in $\ell^1(L)$ defined as $a_n(\ell) = a(\ell n)$. One then easily verify that Φ is a linear isometry between two spaces. Moreover we have

$$\begin{aligned}
(\Phi(a)*\Phi(b)(n)(\ell) &= (\sum_{n_1} \Phi(a)(n_1)\alpha_{n_1}(\Phi(b)(n_1^{-1}n)))(\ell) \\
&= \sum_{\ell_1}\sum_{n_1} \Phi(a)(n_1)(\ell_1)\alpha_{n_1}(\Phi(b)(n_1^{-1}n))(\ell_1^{-1}\ell) \\
&= \sum_{\ell_1\, n_1} a(\ell_1 n_1)b(n_1^{-1}\ell_1^{-1}\ell n) \\
&= (a*b)(\ell n) = \Phi(a*b)(n)(\ell) \quad .
\end{aligned}$$

Hence, $\Phi(a*b) = \Phi(a)*\Phi(b)$ and Φ is an isomorphism. Similarly we can show that $\Phi(a^*) = \Phi(a)^*$. We assert that Φ is an isometry even with respect to the C*-norm $\|a\|_\infty$. Let π be a representation of $\ell^1(M)$, then there exists a unitary representation u of M such that $\pi(a) = \sum_m a(m)u(m)$. Let v and w be the restrictions of u to L and N respectively. Let ρ be the representation of $\ell^1(L)$ associated with v. We see that (ρ, w) is a covariant representation of $(\ell^1(L), N, \alpha)$ such that $\rho \times w \cdot \Phi = \pi$. Indeed,

$$\rho\times w(\Phi(a)) = \sum_{n} \rho(\Phi(a)(n))w(n)$$

$$= \sum_{n,1} a(\ell n)v(\ell)w(n) = \sum_{n,\ell} a(\ell n)u(\ell n) = \pi(a) \quad .$$

Therefore, $\|\Phi(a)\|_\infty \geq \|a\|_\infty$. Conversely, let $\rho\times w$ be a representation of $C^*(L) \rtimes_\alpha N$ (cf. (3.2.1)). Put $u(m) = v(\ell)w(n)$ for an element $m = \ell n \in M$, then u is a unitary representation of M because (ρ, w) is a covariant representation and M is the semidirect product by the action of N. Thus we obtain a representation π of $\ell^1(M)$ through u and as mentioned above $\rho\times w \cdot \Phi = \pi$. It follows that $\|\Phi(a)\|_\infty = \|a\|_\infty$. Therefore, with this embedding Φ we obtain a *-isomorphism of $C^*(M)$ to $C^*(L) \rtimes_\alpha N$.

Now back to the case of the group H we have the identification,

$$C^*(G) = C(\hat{G}) = C(T^2)$$

because $G \cong Z\times Z$. Thus the automorphism α of $C^*(G)$ induces a homeomorphism σ on the 2-torus T^2 as $\alpha(f)(s, t) = f(\sigma^{-1}(s, t))$ $f \in C(T^2)$. We shall determine the map σ. Write an element of G as

$$f_{\ell,n} = \begin{pmatrix} 1 & 0 & \ell \\ 0 & 1 & n \\ 0 & 0 & 1 \end{pmatrix}$$

regarded as the continuous function on T^2 by

$$f_{\ell,n}(s, t) = e^{2\pi i(\ell s + nt)}$$

Then since $\alpha(f_{\ell,n}) = f_{\ell+n,n}$ we have

$$\alpha(f_{\ell,n})(s, t) = f_{\ell+n,n}(s, t)$$

$$= e^{2\pi i((\ell+n)s+nt)} = e^{2\pi i(\ell s+n(s+t))}$$

$$= f_{\ell,n}(s, s+t) = f_{\ell,n}(\sigma^{-1}(s, t)) \quad .$$

Since linear span of the functions $f_{\ell,n}$ forms a dense self-adjoint subalgebra of $C(T^2)$, we see that

$$\sigma^{-1}(s, t) = (s, s+t) \quad , \quad \text{that is} \quad , \quad \sigma(s, t) = (s, t-s) \quad .$$

Thus we can describe the behavior of σ and apply our preceding results to $C^*(H)$.

First of all, the orbit of a point (s, t), $O(s, t)$ sits on the line in T^2 with fixed coordinate s. When s is rational all points on the line $(s, t)(t \in T)$ are periodic points, whereas all points there are not periodic if s is irrational. Thus, in T^2 the set of periodic points for σ is dense so that by (4.1.11) $C^*(H)$ has sufficiently many finite dimensional jrreducible representations. This is however also shown simply by the universal property of $C^*(H)$ and the fact that H is generated by two elements. On the other hand, for a fixed irrational number s the map σ induces an irrational rotation on the torus $\{(s, t) | t \in T\}$. Hence by (2) of (3.3.7) the (pure) state extension to $C^*(H)$ of every evaluation state $\mu_{(s,t)}$ on $C^*(G)$ is unique and it gives rise to an infinite dimensional discrete irreducible representation.

Let u, v and z be unitaries in $C^*(H)$ corresponding to a, b, and c. We can formulate results about $C^*(H)$ in the following way.

Proposition 4.4.2. Take a point (s, t) in T^2 with s being irrational. Then we have;

(1) There exists precisely one state φ on $C^*(H)$ such that

$$\varphi(z) = e^{2\pi is} \quad \text{and} \quad \varphi(v) = e^{2\pi it} \quad .$$

The state φ is pure and has the form $\varphi = \mu_{(s,t)} \cdot \varepsilon$

(2) Let $\pi_{s,t}$ be the GNS-representation of φ on H_φ, then $\{\xi_n = \pi_{s,t}(u)^n \eta(1) | n \in Z\}$ is an orthogonal basis for H_φ. Besides, $\pi_{s,t}(u) = \hat{s}$ and $\pi_{s,t}(v) = e^{2\pi it} d_s$ where $\hat{s}$ is the bilateral shift for $\{\xi_n\}$ and d_s is the diagonal

operator given by $d_s\xi_n = e^{-2\pi ins}\xi_n$.

Proof. Let φ be a state on $C^*(H)$ such that

$$\varphi(z) = e^{2\pi is} \quad \text{and} \quad \varphi(v) = e^{2\pi it} \quad .$$

We assert first that

$$\varphi(az) = \varphi(a)\varphi(z) \quad \text{and} \quad \varphi(av) = \varphi(a)\varphi(v) \qquad a \in C^*(H) \quad .$$

In fact, let ξ_φ be the canonical cyclic vector in H_φ. We have then

$$1 = |(\pi_{s,t}(z)\xi_\varphi, \xi_\varphi)| \le \| \pi_{s,t}(z)\xi_\varphi\| \, \|\xi_\varphi\| \le 1 \quad .$$

Hence,

$$|(\pi_{s,t}(z)\xi_\varphi, \xi_\varphi)| = \| \pi_{s,t}(z)\xi_\varphi\| \, \|\xi_\varphi\|$$

so that $\pi_{s,t}(z)\xi_\varphi = \lambda\xi_\varphi$ and necessarily $\lambda = \varphi(z)$. It follows that

$$\begin{aligned} \varphi(az) &= (\pi_{s,t}(az)\xi_\varphi, \xi_\varphi) \\ &= (\pi_{s,t}(a)\pi_{s,t}(z)\xi_\varphi, \xi_\varphi) \\ &= \lambda(\pi_{s,t}(a)\xi_\varphi, \xi_\varphi) = \varphi(a)\varphi(z) \quad . \end{aligned}$$

Similarly, $\varphi(av) = \varphi(a)\varphi(v)$. Now $C(T^2)$ is generated by $z = f_{1,0}$ and $v = f_{0,1}$ and this means that $\varphi|C(T^2)$ coincides with the evaluation $\mu_{(s,t)}$ at the point (s, t). Thus, as mentioned before, φ is a unique pure state extension of $\mu_{(s,t)}$.

The assertion (2) easily follows from the structure of $\pi_{s,t}$ described in Secs. 3.3 and 4.1 as an induced representation of the trivial isotropy subgroup $H_{(s,t)} = \{0\}$. For instance,

$$\begin{aligned} \pi_{s,t}(v)\xi_n &= f_{0,1}(\sigma^n(s, t))\xi_n \\ &= e^{2\pi i(t - ns)}\xi_n = e^{2\pi it}d_s\xi_n \quad . \end{aligned}$$

Let π be an irreducible representation of $C^*(H)$ on a Hilbert space K. Then as z is central there is a number $s(\pi)$ of T such that $\pi(z) = e^{2\pi i s(\pi)}1$, namely

$$\pi(u)\pi(v) = e^{2\pi i s(\pi)}\pi(v)\pi(u) \quad .$$

Thus, the C^*-algebra $\pi(C^*(H))$ is generated by two unitaries $\pi(u)$ and $\pi(v)$ satisfying the above commutation relation. Therefore, if $s(\pi)$ is irrational it coincides with the irrational rotation C^*-algebra $A_{s(\pi)}$. If two irreducible representations π_1 and π_2 are equivalent we must have $s(\pi_1) = s(\pi_2)$. Hence the irreducible representations of $C^*(H)$ are in one-to-one correspondence with the irreducible representations of the irrational and the rational rotation algebras.

The definition of a discrete irreducible representation $\{\pi, K\}$ is reduced here to say that $\pi(v)$ has an eigenvalue. Now suppose that such a representation π with $s(\pi)$ being irrational has an eigenvalue $e^{2\pi it}$ for $\pi(v)$. Let $\xi_0 \in K$ be a unit eigenvector and define the state φ on $C^*(H)$ by

$$\varphi(a) = (\pi(a)\xi_0, \xi_0) \quad .$$

Then,

$$\varphi(z) = e^{2\pi i s(\pi)} \quad \text{and} \quad \varphi(v) = e^{2\pi it} \quad ,$$

whence by the above proposition φ is nothing but the unique pure state extension of the evaluation state at the point $(s(\pi), t)$ and π is equivalent to the representation $\pi_{(s(\pi),t)}$. One can also verify that the eigenvalues of $\pi(v)$ consists precisely of $\{e^{2\pi i(t - ns(\pi))} | n \in Z\}$. Moreover, the isotropy subgroup $H_{(s(\pi),t)}$ is trivial. Hence (4.1.4) immediately tells us that

> "Two infinite dimensional discrete irreducible representations of $C^*(H)$, π_1 and π_2 with their associated points (s_1, t_1) and (s_2, t_2) are unitarily equivalent if and only if $O(s_1, t_1) = O(s_2, t_2)$".

We remark that this result also indicates same conclusion for the unitary equivalence of (infinite dimensional) irreducible discrete representations of the irrational rotation algebra A_s.

A finite dimensional irreducible representation of C*(H) arises from a periodic point (s, t) for a rational number s, the period of which depends only on s. The set of irreducible representations of a fixed dimension n has the structure described in Sec. 4.2. We leave the reader detailed analysis of finite dimensional case.

Finally we note that C*(G) is a maximal abelian subalgebra of C*(H) by (3.3.2) (or (4.3.5)). Indeed, here every set T^2_n has no interior because

$$(T^2)^n = \{(s, t) \in T^2 | \sigma^n(s, t) = (s, t)\}$$

$$= \{(\frac{k}{n}, t | k = 0,1,2,\ldots,n-1\}$$

CHAPTER 5

SHIFT DYNAMICAL SYSTEMS AND TOPOLOGICAL DYNAMICAL SYSTEMS ASSOCIATED WITH SUBGROUPS OF THE TORUS T

In this chapter we first introduce shift dynamical systems which are closely related to the families of shift operators on a separable Hilbert space and discuss their structure. The next section is devoted to introduce the notion of K-groups of C*-algebras together with relevant properties of them for applications. Readers who are interested in further details of the K-theory of C*-algebras may consult with the book [A]. Results in both sections will be used in the last section to give the complete algebraic invariant of transformation group C*-algebras associated with those dynamical systems arising from infinite subgroups of T_d (the torus with discrete topology).

5.1 Shift Dynamical Systems

Throughout this chapter we fix a separable infinite dimensional Hilbert space H with a fixed complete orthonormal base $\{\xi_n | n \in Z\}$. We denote $\mathscr{D}$ the diagonal algebra for $\{\xi_n\}$ which consists of all operators a's such that $a\xi_n = \lambda_n \xi_n$ for $(\lambda_n) \in \ell^\infty(Z)$. Thus, in this sense we identify $\mathscr{D}$ with $\ell^\infty(Z)$. It is a maximal abelian von Neumann subalgebra of $B(H)$ generated by minimal projections $\{p_n\}$ for vectors $\{\xi_n\}$. We write s the simple bilateral shift operator for $\{\xi_n\}$, that is, $s\xi_n = \xi_{n+1}$. On the contrary, when the base $\{\xi_n\}$ is indexed by the natural number N, the operator s is called the unilateral shift operator. By a shift operator u we mean a unitary operator such that

$$u\xi_n = \lambda_{n+1}\xi_{n+1} \qquad \text{with} \qquad |\lambda_{n+1}| = 1 \quad .$$

The operator u is written as $u = ws$ where w is a diagonal unitary operator such that $w\xi_n = \lambda_n\xi_n$. Moreover, it is unitarily equivalent to the simple shift operator s. Indeed, if we define the unitary operator v by

$$v\xi_n = \lambda_n \cdot \lambda_{n-1}, \ldots, \lambda_1\xi_n \qquad n \geq 1$$

$$v\xi_0 = \xi_0$$

$$v\xi_n = \bar{\lambda}_{n+1} \cdot \bar{\lambda}_n \ldots \bar{\lambda}_0\xi_n \qquad n \leq -1 \quad ,$$

we obtain the identity $s = v^*uv$.

There are many results in literature about invariant subspaces (both one-sided and two-sided) for shift operators. The starting Beurling theorem (1949) concerned with the structure of invariant subspaces of $L^2(T)$ for the algebra $H^\infty(T)$ (one-sided invariant subspace problem) where $H^\infty(T)$ means the $\sigma(L^\infty(T), L^1(T))$-closed subalgebra of $L^\infty(T)$ generated by the function $z : z(e^{2\pi is}) = e^{2\pi is}$. Invariant subspace problems with arbitrary multiplicity have been treated recently by Kawamura[16].

Definition 5.1.1. A shift dynamical system $\Sigma = (X, \sigma, \phi)$ is a topological dynamical system (X, σ) with a map $\phi : Z \to X$ such that $\sigma(\phi(n)) = \phi(n+1)$ and the image $\phi(Z)$ is dense in X.

By definition a shift dynamical system becomes a topologically transitive dynamical system in which a dense orbit $O(\phi(0))$ is specified.

Let $\mathscr{S}$ be a family of shift operators on H and $W(\mathscr{S})$ be the diagonal family for $\mathscr{S}$, that is,

$$w(\mathscr{S}) = \{w | u = ws \in \mathscr{S}\} \quad .$$

As mentioned before, a family $\mathscr{S}$ is unitarily equivalent to the family of shift operators containing the operator s. Thus we consider the family $\mathscr{S}$ satisfying the following conditions:

(a) $s \in \mathscr{S}$

(b) $sW(\mathscr{S})s^* = W(\mathscr{S})$

(c) $W(\mathscr{S})$ is a group.

Note that to consider (two-sided) invariant subspaces for $\mathscr{S}$ is equivalent to consider invariant subspaces for the C*-algebra $C^*(\mathscr{S})$ generated by $\mathscr{S}$. The latter C*-algebra is generated by the family $W(\mathscr{S})$ and s, and ads acts on the commutative C*-algebra $A = C^*(W(\mathscr{S}))$ as a *-automorphism. We do not, however, intend to study the invariant subspace here.

Let $C(X)$ be the Gelfand representation algebra of A. The adjoint ads induces a *-automorphism α on $C(X)$ together with a homeomorphism σ on X. Define a character $\phi(n)$ on $C(X)$ (i.e. a point of X) by $\phi(n)(\hat{a}) = \lambda_n$ for $a = \sum_{k \in Z} \lambda_k p_k \in A$. Then the system $\Sigma = (X, \sigma, \phi)$ becomes a shift dynamical system. Indeed, since

$$\begin{aligned}\sigma(\phi(n))(a) &= \phi(n)(\alpha^{-1}(a)) \\ &= \phi(n)(\widehat{s^*as}) = \lambda_{n+1} = \phi(n+1)(\hat{a}) \quad ,\end{aligned}$$

we have $\sigma(\phi(n)) = \phi(n+1)$ and moreover as the Gelfand representation of A is an isometry the image $\phi(Z)$ is dense in X. Namely the above family of shift operators $\mathscr{S}$ gives rise to a shift dynamical system Σ. Next suppose that a shift dynamical system $\Sigma = (X, \sigma, \phi)$ is given. We define the representation π_ϕ of $C(X)$ on H by $\pi_\phi(f)\xi_n = f(\phi(n))\xi_n$. Apparently, π_ϕ is faithful and $\pi_\phi(C(X)) \subset \mathscr{D}$. Furthermore, we have $\pi_\phi(\alpha(f)) = s\pi_\phi(f)s^*$, that is, $\{\pi_\phi(C(X)), s^n, Z\}$ is a covariant representation of the action $\{C(X), Z, \alpha\}$. Therefore, the system Σ gives rise to a family $\mathscr{S}$ of shift operators where

$$\mathscr{S} = \{u = ws \mid w \text{ is a unitary operator of } \pi_\phi(C(X))\} \quad .$$

Thus we have seen the correspondence between families of shift operators (satisfying the conditions (a), (b) and (c)) and shift dynamical systems.

Henceforth, we write $C^*(\Sigma)$ instead of $C^*(\mathscr{S})$ as the C*-algebra associated with a shift dynamical system in the above way.

Now consider first the case where X consists of finite points, say

$$X = \{x_0, x_1, x_2, \ldots, x_{n-1}\} \quad ,$$

in which σ acts on X as a cyclic shift. In this case $\pi_\phi(C(X))$ is a C*-algebra spanned by n orthogonal projections:

$$e_i = \sum_{k \in Z} p_{nk+i} \qquad (i = 0,1,\ldots,n-1) \quad .$$

Define the unitary operator $v : H \to H \otimes H_n$ by

$$v(\xi_{nk+i}) = \xi_k \otimes \eta_i \qquad i = 0,1,\ldots,n-1 \quad , \quad k \in Z \quad ,$$

where $\{\eta_i\}$ is an orthonormal base for an n-dimensional Hilbert space H_n. We then see that

$$vC^*(\Sigma)v^* = C^*(s) \otimes B(H_n) = M_n(C^*(s)) \quad .$$

Since the spectrum of s fills up the whole unit circle, $C^*(\Sigma)$ is *-isomorphic to the $n \times n$ block matrix algebra $M_n(C(S^1))$ over the algebra $C(S^1)$), so that we need no further investigation about the structure of the algebra $C^*(\Sigma)$.

Proposition 5.1.1. For a shift dynamical system $\Sigma = (X, \sigma, \phi)$ the following conditions are equivalent;

(1) X is an infinite set;
(2) The map ϕ is injective;
(3) $\phi(Z)$ is a proper subset of X.

Proof. The assertion $(1) \Rightarrow (3)$. Suppose that $\phi(Z) = X$, then the category theorem tells us that there exists an isolated point $\phi(n_0)$. It follows that every point $\phi(n)$ becomes an isolated point, a contradiction. Assume next the condition (3) and suppose ϕ were not injective. There exists then a pair of integers (m_0, n_0) with $n_0 > m_0$ such that $\phi(n_0) = \phi(m_0)$. Put $k = n_0 - m_0$, then

$$\phi(n) = \sigma^{n-m_0}(\phi(m_0)) = \sigma^{n-m_0}(\phi(n_0)) = \phi(n+k)$$

for every $n \in Z$. Hence the image $\phi(Z)$ consists of at most k-points and $\phi(Z) = X$, a contradiction. As the implication, $(2) \Rightarrow (1)$ is trivial, this completes the proof.

We assume henceforth that X is always an *infinite set*. Note that in this case we have the following

Proposition 5.1.2. $C^*(\Sigma)$ is irreducible on H.

Proof. Take an operator a commuting with the algebra $C^*(\Sigma)$. For any pair (i, j) with $i \neq j$, there exists, by (2) of the above proposition, a function $f \in C(X)$ such that $f(\phi(i)) = 1$ and $f(\phi(j)) = 0$. Then

$$(a\xi_i, \xi_j) = (a\pi_\phi(f)\xi_i, \xi_j) = (a\xi_i, \pi_\phi(f)^*\xi_j) = 0 \quad ,$$

which means that a is of the form $\sum_{n \in Z} \lambda_n p_n$. Besides, a commutes with the shift operator s. Therefore, $a = \lambda 1$ for some scalar λ.

Among the C*-algebras associated with shift dynamical systems, the particular algebra is the one associated with the full shift dynamical system $\Sigma_0 = (\beta Z, \sigma, \phi)$ where σ is the extension of the simple shift on the integers Z. Here, $\pi_\phi(C(\beta Z)) = \mathcal{D}$ and as mentioned before (Sec. 3.3) the system Σ_0 has no periodic points.

In the algebra $B(H)$, there is a faithful normal projection of norm one from $B(H)$ to the diagonal algebra $\mathcal{D}$ defined as

$$x \in B(H) \to \sum_n p_n x p_n \in \mathscr{D} \quad .$$

We denote its restriction to $C^*(\Sigma)$ by ε_Σ. By definition,

$$\varepsilon_\Sigma(\pi_\phi(f)s^k) = 0 \quad \text{if} \quad k \neq 0 \quad ,$$

hence

$$\varepsilon_\Sigma\Big(\sum_{-n}^{n} \pi_\phi(f_k)s^k\Big) = \pi_\phi(f_0) \quad .$$

Since $C^*(\Sigma)$ is the closure of those elements in the above form, we have

$$A = \varepsilon_\Sigma(C^*(\Sigma)) = C^*(\Sigma) \cap \mathscr{D} \quad .$$

Namely, ε_Σ is a faithful (normal) projection of norm one of $C^*(\Sigma)$ to $A = \pi_\phi(C(X))$.

Proposition 5.1.3. Let $\Sigma = (X, \sigma, \phi)$ be a shift dynamical system. Then

(1) $C^*(\Sigma)$ is *-isomorphic to the crossed product $C(X) \rtimes_\alpha Z$.
(2) If (X, σ) is minimal, $C^*(\Sigma)$ is a simple C*-algebra with a faithful tracial state $\tau = \varphi \cdot \varepsilon_\Sigma$ for a faithful α-invariant state φ on $\pi_\phi(C(X))$. Every tracial state of $C^*(\Sigma)$ is of this form.

Proof. Let ρ be the canonical homoemorphism of $C(X) \rtimes_\alpha Z$ to $C^*(\Sigma)$ associated with the covariant representation $\{\pi_\phi, s^n\}$. We then have the relation:

$$\varepsilon_\Sigma \cdot \rho(a) = \rho \cdot \varepsilon(a) \qquad a \in C(X) \rtimes_\alpha Z \quad ,$$

the same situation as in (4.3.1). Hence, ρ is an isomorphism. The second assertion is an immediate consequence of (4.3.3) and (3.2.9) whereas the existence of a faithful α-invariant state on $\pi_\phi(C(X))$ is assured by (1.1.5).

Now consider two shift dynamical systems $\Sigma_1 = (X_1, \sigma_1, \phi_1)$ and $\Sigma_2 = (X_2, \sigma_2, \phi_2)$, and suppose that they are topologically conjugate. There exists then a *-isomorphism θ of $C(X_1)$ to $C(X_2)$ such that $\theta \cdot \alpha_1 = \alpha_2 \cdot \theta$ where α_i is the associated *-automorphism of $C(X_i)$. It follows by (3.2.2) that $C(X_1) \rtimes_{\alpha_1} Z$ is isomorphic to $C(X_2) \rtimes_{\alpha_2} Z$. Namely, $C^*(\Sigma_1)$ is *-isomorphic to $C^*(\Sigma_2)$ (cf. 3.2.3)). It is however to be noticed that even if $C^*(\Sigma_1)$ is spatially isomorphic to $C^*(\Sigma_2)$, Σ_1 is not necessarily topologically conjugate to Σ_2 as we see from the next exmaple.

<u>Example 5.1.1.</u> Put $X = \{w_1, w_2\} \cup Z \cup \{w_3\}$, where w_1 (resp. w_2) is the limit point of the set of positive even integers (resp. positive odd integers) and w_3 is the limit point of the set of negative integers. Let σ_1 and σ_2 be the homeomorphisms of X such that

$$\sigma_1(n) = n + 1 \quad , \quad \sigma_2(n) = n - 1 \quad n \in Z$$

$$\sigma_i(w_1) = w_2 \quad , \quad \sigma_i(w_2) = w_1 \quad , \quad \sigma_i(w_3) = w_3 \quad .$$

Let ϕ_1 and ϕ_2 be the maps of Z into X such that $\phi_1(n) = n$ and $\phi_2(n) = -n$. The systems $\Sigma_1 = (X, \sigma_1, \phi_1)$ and $\Sigma_2 = (X, \sigma_2, \phi_2)$ are then shift dynamical systems. We see that $vC^*(\Sigma_1)v^* = C^*(\Sigma_2)$ by the unitary operator v defined as $v\xi_n = \xi_{-n}$. However, there exists no homeomorphism σ of X such that $\sigma \cdot \sigma_1 = \sigma_2 \cdot \sigma$.

Our present subject is not directly concerned with the determination of the conjugate classes of shift dynamical systems. We remark however that the conjugate classes of such dynamical systems are known to be corresponding to the isomorphic classes of analytic Banach algebras together with their diagonal parts provided that the homeomorphisms admit ergodic invariant measures with full support sets, where those Banach algebras mean the closures of those elements,

$$\sum_{k=0}^{n} \pi_\phi(f_k)s^k.$$

<u>Definition 5.1.2.</u> Two shift dynamical systems $\Sigma_i = (X_i, \sigma_i, \phi_i)$

$(i = 1,2)$ are said to be strictly conjugate if there exists a homeomorphism $h : X_1 \to X_2$ such that $h \cdot \phi_1(n) = \phi_2(n)$.

Strictly conjugate systems are topologically conjugate. It should be noticed that the same dynamical system could define different shift dynamical systems which are not strictly conjugate. In fact, consider the Bernouill system $X(k) = (\pi S_k, \sigma_k)$ for k-symbols. Let x_0 be the point of $X(k)$ whose components contain the parts of k-adic expansion of all positive integers on both sides of zero component. Let x_1 be the point of $X(k)$ whose components contain the same ones as in those of x_0 but only on the right side of zero component. Then by definition, $\sigma_k^n(x_1)$ converges to zero as $n \to \infty$, whereas $\sigma_k^n(x_0)$ does not converge to zero. Therefore, the system $(X(k), \sigma_k, x_0)$ is not strictly conjugate to the system $(X(k), \sigma_k, x_1)$.

Proposition 5.1.4. Two shift dynamical systems $\Sigma_i = (X_i, \sigma_i, \phi_i)$ $(i = 1,2)$ are strictly conjugate if and only if $\pi_{\phi_1}(C(X_1)) = \pi_{\phi_2}(C(X_2))$.

Proof. Let h be a homeomorphism of X_1 to X_2 such that $h \cdot \phi_1(n) = \phi_2(n)$ for every $n \in Z$. For every function $f \in C(X_2)$ we have $g = f \cdot h \in C(X_1)$ and

$$\begin{aligned} \pi_{\phi_1}(g)\xi_n &= f \cdot h(\phi_1(n))\xi_n = f(\phi_2(n))\xi_n \\ &= \pi_{\phi_2}(f)\xi_n \quad . \end{aligned}$$

Hence, $\pi_{\phi_2}(g) = \pi_{\phi_1}(f)$ which implies that $\pi_{\phi_1}(C(X_1)) = \pi_{\phi_2}(C(X_2))$. Conversely, if $\pi_{\phi_1}(C(X_1)) = \pi_{\phi_2}(C(X_2))$ the map $\theta = \pi_{\phi_1}^{-1} \cdot \pi_{\phi_2}$ is a *-isomorphism of $C(X_2)$ to $C(X_1)$. Hence there exists a homeomorphism $h : X_1 \to X_2$ such that $\theta(f)(x) = f(h(x))$ $f \in C(X_2)$. One then sees that

$$\begin{aligned} f(h(\phi_1(n)))\xi_n &= \pi_{\phi_1}(\theta(f))\xi_n = \pi_{\phi_2}(f)\xi_n \\ &= f(\phi_2(n))\xi_n \qquad f \in C(X_2) \quad . \end{aligned}$$

Hence $h(\phi_1(n)) = \phi_2(n)$. This completes the proof.

Recall that a compact abelian infinite group Γ is called a monothetic group if there exists an injective homomorphism ϕ of Z into Γ such that $\phi(Z)$ is dense in Γ (cf. Appendix A). In this case, considering the homeomorphism σ of Γ defined by $\sigma(x) = x + \phi(1)$ we obtain a shift dynamical system $\Sigma = (\Gamma, \sigma, \phi)$. By definition, the family $\{\sigma^n | n \in Z\}$ is an equicontinuous family of homeomorphisms and the system is minimal. We shall make use of these facts in the last section.

Proposition 5.1.5. Suppose that a shift dynamical system $\Sigma = (X, \sigma, \phi)$ satisfies either of the following conditions:

(1) $\phi(n)$ is an isolated point for some n (hence for all n);
(2) Σ coincides with the system induced by a monothetic group.

Then a shift dynamical system $\Sigma_0 = (X_0, \sigma_0, \phi_0)$ is strictly conjugate to Σ if and only if it is conjugate to Σ.

Proof. The proof for the condition (1). Let $k : X \to X_0$ be a homeomorphism with $k \cdot \sigma = \sigma_0 \cdot k$. Write

$$X_0 = k(\phi(Z)) \cup k(X \sim \phi(Z)) \quad .$$

Since, by the assumption, $\phi(Z)$ is an open set, $k(X \sim \phi(Z))$ is a σ_0-invariant closed set. Hence $\phi_0(0)$ must belong to $k(\phi(Z))$ and there exists an integer m such that $\phi_0(0) = k(\phi(m))$. Put $h = \sigma_0^m \cdot k$. We then have that

$$h(\phi(n)) = \sigma_0^m \cdot k(\phi(n))$$

$$= \sigma_0^m \cdot k \cdot \sigma^{n-m}\phi(m) = \phi_0(n) \quad , \quad n \in Z \quad .$$

The proof for the condition (2). Let k be the same as above and define the homeomorphism h of X to X_0 as

$$h(x) = k(x + k^{-1}(\phi_0(0))) \quad .$$

Then for every n

$$h(\phi(n)) = k(\phi(n) + k^{-1}(\phi_0(0))) = k \cdot \sigma^n k^{-1}(\phi_0(0))$$

$$= k \cdot k^{-1} \cdot \sigma_0^n(\phi_0(0)) = \phi_0(n) \quad .$$

For two shift dynamical systems Σ_1 and Σ_2, we see that $C^*(\Sigma_1) = C^*(\Sigma_2)$ if and only if $\pi_{\phi_1}(C(X_1)) = \pi_{\phi_2}(C(X_2))$. Therefore by (5.1.4) we have the following

Corollary 5.1.6. Assume the same condition for a shift dynamical system Σ as in (5.1.5), then Σ_0 is conjugate to Σ if and only if $C^*(\Sigma_0) = C^*(\Sigma)$.

As stated before, the C*-algebra $C^*(\Sigma)$ is irreducible on H when X is an infinite set. For an irreducible C*-algebra B on H it is a deep result in the theory of C*-algebra that either $B \supset C(H)$ or $B \cap C(H) = \{0\}$. With this result we can prove the following

Theorem 5.1.7. For a shift dynamical system $\Sigma = (X, \sigma, \phi)$ the following conditions are equivalent:

(1) $C^*(\Sigma) \supset C(H)$;

(2) $\phi(Z)$ is open in X;

(3) $\phi(n)$ is an isolated point for some n (condition (1) of (5.1.4)).

Proof. (1) $\Rightarrow$ (3). Suppose that $C^*(\Sigma) \supset C(H)$, then $p_n \in C^*(\Sigma)$ and

$$p_n \in C^*(\Sigma) \cap \mathscr{D} = \pi_\phi(C(X)) \quad .$$

Hence, $p_n = \pi_\phi(f)$ for some function f. From the definition of the representation π_ϕ one verifies easily that f is the characteristic function of the set $\{\phi(n)\}$. Hence $\phi(n)$ is an isolated point.

(2) $\Rightarrow$ (1). Take a neighborhood U of $\phi(0)$ such that $\bar{U} \subset \phi(Z)$. Since $\phi(Z)$ is a countable set this implies by the category theorem that there exists an isolated point $\phi(n_0)$ in $\bar{U}$. The characteristic

function f of the set $\{\phi(n_0)\}$ is then continuous and $\pi_\phi(f) = p_{n_0}$. Hence, $C^*(\Sigma) \cap C(H) \neq \{0\}$ and by the above cited theorem we have the assertion (1). As the implication (3) $\Rightarrow$ (2) is trivial, this completes the proof.

Examples of shift dynamical systems where $\phi(Z)$ is open in X are illustrated as $X = \beta Z$, one point compactification of Z and $Z \cup \{w_0, w_1, w_2, \ldots, w_{p-1}\}$ etc. In the last case, each w_i is the limit point of the set $\{pZ + i\}$ and the homeomorphism is specified as the map: $n \to n+1$ on Z and the cyclic map on the set $\{w_i\}$.

Now with the condition for $\phi(Z)$ set $X' = X \sim \phi(Z)$. Then X' is an invariant closed set and we may consider the dynamical system $\Sigma' = (X', \sigma|X')$. On the other hand, let $I_0 = k(X')$, and let I be the ideal of $C^*(\Sigma)$ generated by $\pi_\phi(I_0)$ and s. In this case it is not so hard to see that $I = C(H)$. Furthermore, when Σ' becomes also a shift dynamical system we can prove the reduction that the quotient algebra $C^*(\Sigma)/C(H) = C^*(\Sigma)/I$ is *-isomorphic to the algebra $C^*(\Sigma')$.

5.2 K-Groups of C*-algebras and Topological Dynamical Systems

In this section we introduce the K-groups of C*-algebras which have been playing very important rôles in the theory of C*-algebras. We begin with the construction of the universal group of a semigroup. Let S be an abelian semigroup. There is then a canonical way of constructing an abelian group $\mathscr{U}(S)$ (called the universal or Grothendieck group of S). Namely in the product space $S \times S$ we set a relation $(a, b) \sim (c, d)$ if there exists an element $r \in S$ such that

$$a + d + r = b + c + r \quad .$$

One then easily sees that this is an equivalence relation. Let $\mathscr{U}(S)$ be the set of equivalence classes of pairs (a, b), then $\mathscr{U}(S)$ becomes a group under the operation induced by that of S. The diagonal set $\Delta(S)$ of S in $S \times S$ forms a single equivalence class in $\mathscr{U}(S)$ as the identity element since $(a + r, b + r) \sim (a, b)$. The inverse of the class $[(a, b)]$ is $[(b, a)]$. We note that if S has the cancellation property the equivalence $(a, b) \sim (c, d)$ simply reduces to the identity: $a + d = b + c$. For $a \in S$ the map α_S

$$\alpha_S : a \to [(a, 0)] \in \mathcal{U}(S)$$

defines the embedding of S into $\mathcal{U}(S)$. If S is a group, α_S is clearly a group isomorphism. Now, by definition, one can further verify that if γ is a homomorphism of a semigroup S_1 to a semigroup S_2 there is an associate group homomorphism γ_* of $\mathcal{U}(S_1)$ to $\mathcal{U}(S_2)$ such that the following diagram is commutative;

$$\begin{array}{ccc} S_1 & \xrightarrow{\alpha_{S_1}} & (S_1) \\ \gamma \downarrow & & \downarrow \gamma_* \\ S_2 & \xrightarrow{\alpha_{S_2}} & (S_2) \end{array}$$

This construction will be used to define the K_0-group of a C*-algebra for the semigroup of the equivalence classes of projections.

Definition 5.2.1. Let e and f be idempotents in a C*-algebra A. We write;

(a) $e \sim f$ (algebraic equivalence) if there are $x, y \in A$ with $xy = e$ and $yx = f$;

(b) $e \underset{s}{\sim} f$ (similar) if there is an invertible element z with $zez^{-1} = f$;

(c) $e \underset{h}{\sim} f$ (homotopic) if there is a norm continuous path of idempotents in A from e to f.

Two projections p and q are said to be equivalent as projections if there is an element v (called a partial isometry) in A such that $v^*v = e$ and $vv^* = f$. If v is chosen as a unitary we say that they are unitarily equivalent.

In order to define and understand the concept of K-groups of C*-

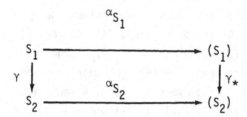

algebras, we shall investigate relations among these equivalences but we must recognize first the following fact.

Lemma 5.2.1. $e \sim f$ is an equivalence relation.

Proof. Suppose that $e \sim f$ with x, y and $f \sim g$ with z, w. Then

$$(exfzg)(gwfye) = exfzwzwfye$$
$$= exfye = e^4 = e \quad ,$$

and similarly

$$(gwfye)(exfzg) = g \quad .$$

These calculations show that in the relation $e \sim f$ we may usually assume as $x = ex = xf$, $y = fy = ye$. Henceforth we keep this remark.

Lemma 5.2.2. If $e \sim f$, then $\begin{pmatrix} e & 0 \\ 0 & 0 \end{pmatrix} \underset{s}{\sim} \begin{pmatrix} f & 0 \\ 0 & 0 \end{pmatrix}$ in $M_2(A)$.

Proof. Take two elements x, y giving the equivalence $e \sim f$ and put $z = \begin{pmatrix} y & 1-f \\ 1-e & x \end{pmatrix}$. Then $z^{-1} = \begin{pmatrix} x & 1-e \\ 1-f & y \end{pmatrix}$ and

$$z\begin{pmatrix} e & 0 \\ 0 & 0 \end{pmatrix}z^{-1} = \begin{pmatrix} f & 0 \\ 0 & 0 \end{pmatrix} .$$

Let $GL_n(A)$ (resp. $U_n(A)$) be the group of invertible elements (resp. unitaries) in $M_n(A)$. In the following all homotopic path mean norm continuous paths.

Proposition 5.2.3. Let x and y be invertible elements of A, then there is a path in $GL_2(A)$ from $\mathrm{diag}(xy, 1)$ to $\mathrm{diag}(x, y)$. If x and y are unitaries, the path may be chosen in $U_2(A)$. In particular, for an invertible element z there is a path in $GL_2(A)$ from 1 to $\mathrm{diag}(z, z^{-1})$.

For the proof it is enough to consider the path

$$z_t = \mathrm{diag}(x, 1)u_t\mathrm{diag}(y, 1)u_t^* \quad ,$$

where

$$u_t = \begin{pmatrix} \cos\frac{\pi}{2}t & -\sin\frac{\pi}{2}t \\ \sin\frac{\pi}{2}t & \cos\frac{\pi}{2}t \end{pmatrix} .$$

Since $\mathrm{diag}(x, y)$ and $\mathrm{diag}(y, x)$ are connected by the path $u_t\mathrm{diag}(x, y)u_t^*$, the above result tells us that $\mathrm{diag}(xy, 1)$ is homotopic to $\mathrm{diag}(yx, 1)$ in $GL_2(A)$. This concerns the commutativity of the K_1-group of C*-algebras.

<u>Lemma 5.2.4.</u> If $e \underset{s}{\sim} f$, then $\begin{pmatrix} e & 0 \\ 0 & 0 \end{pmatrix} \underset{h}{\sim} \begin{pmatrix} f & 0 \\ 0 & 0 \end{pmatrix}$.

<u>Proof.</u> Let $zez^{-1} = f$ and consider the path w_t in the above proposition connecting 1 and $\mathrm{diag}(z, z^{-1})$. Then $e_t = w_t\mathrm{diag}(e, 0)w_t^{-1}$ provides a homotopy path between $\begin{pmatrix} e & 0 \\ 0 & 0 \end{pmatrix}$ and $\begin{pmatrix} f & 0 \\ 0 & 0 \end{pmatrix}$.

In general, $e \underset{s}{\sim} f$ does not necessarily imply the homotopy equivalence, $e \underset{h}{\sim} f$.

So far we have observed equivalences among idempotents in a C*-algebra, but they will be essentially reduced to relations among projections.

<u>Proposition 5.2.5.</u> Every idempotent e in A is homotopic to a projection p by the path of the form $p_t = w_t e w_t^{-1}$. In particular, $e \underset{s}{\sim} p$.

<u>Proof.</u> Put $z = 1 + (e - e^*)(e^* - e)$. Then z is an invertible positive element and $ze = ez = ee^*e$. Hence its inverse r commutes with e and e^*. Set $p = ee^*r$, then by the above properties of z and r we see that p is a projection with $ep = p$, $pe = e$. Therefore, the element $w_t = 1 - tp + te$ is invertible for any real t with inverse $1 - te + tp$. Hence we obtain a homotopy path from e to p by $p_t = w_t e w_t^{-1}$.

We remark that a parametrization of the above argument shows that if $p \underset{h}{\sim} q$ the homotopy path may be chosen to consist of projections.

Now an immediate consequence of this projection is that if $e \sim f$ their corresponding projections p and q are equivalent. Moreover we have

Proposition 5.2.6. If $p \sim q$, then they are equivalent as projections. Furthermore if $p \underset{s}{\sim} q$, they are unitarily equivalent.

Proof. Let $p = xy$ and $q = yx$. Then $p = y^*x^*xy \le \|x\|^2 y^*y$, and y^*y is invertible in the algebra pAp. Similarly yy^* is invertible in qAq. Let r be the inverse of $(y^*y)^{\frac{1}{2}}$ in pAp and set $v = yr$. Then $v^*v = ry^*yr = p$. On the other hand, $vv^* \le q$ and

$$(q - vv^*)y = y - yr^2y^*y = y - yp = 0 \quad .$$

As yy^* is invertible in qAq, we have $vv^* = q$. Next, if $zpz^{-1} = q$ then $u = z(z^*z)^{-\frac{1}{2}}$ is a unitary and $upu^* = q$. Indeed, $zp = qz$ implies $pz^* = z^*q$ whence p commutes with z^*z and with $(z^*z)^{\frac{1}{2}}$. It follows that $upu^* = q$.

The above arguments mean that we have made use of the polar decomposition of y and z within the C*-algebra A. These devices are often used in the discussions of the K-theory of C*-algebras.

Lemma 5.2.7. Let p and q be projections in A. If $\|p - q\| < 1$, then p and q are unitarily equivalent.

Proof. Put $x = qp$, then x^*x is invertible in pAp because

$$\|x^*x - p\| = \|p(q - p)p\| < 1 \quad .$$

Set $v = xy^{\frac{1}{2}}$ where y is the inverse of x^*x in pAp. Then $v^*v = y^{\frac{1}{2}}x^*xy^{\frac{1}{2}} = p$. On the other hand, $vv^* \le q$ and

$$(q - vv^*)x = x - xyx^*x = x - xp = 0 \quad .$$

It follows that $q = vv^*$ for xx^* is invertible in qAq. Apply the same argument for $1-p$ and $1-q$ and find the partial isometry v' for the equivalence $1-p \sim 1-q$. Then the unitary $u = v + v'$ provides the unitary equivalence, $upu^* = q$.

As a consequence we have the following

Proposition 5.2.8. If $p \underset{h}{\sim} q$ with the path p_t of projections, then there is a path u_t of unitaries such that $u_0 = 1$ and $u_t p u_t = p_t$. In particular, p and q are unitarily equivalent.

Proof. Let $t_0 = 0 < t_1 < t_2 < \ldots < t_n = 1$ be a decomposition of the interval $[0, 1]$ so that $\| p_t - p_s \| < 1$ if s and t are in the same interval. We apply the preceding argument for the elements $x_t = p_t p$ and $x_t' = (1-p_t)(1-p)$ for $t \in [0, t_1]$, and get the path u_t of unitaries with $u_0 = 1$ and $u_{t_1} p u_{t_1}^* = p_{t_1}$. Similarly, we can find a path u_t of unitaries on each interval $[t_i, t_{i+1}]$ connecting p_{t_i} and p_t. It follows that the path $\hat{u}_t = u_t \cdot u_{t_i} \ldots u_{t_1}$ provides the homotopy from p to p_t.

Now we shall define the K_0-group of a unital C*-algebra A. Let $\mathrm{Proj}_n(A)$ be the projections in $M_n(A)$. We have then natural sequence of embeddings;

$$\mathrm{Proj}_1(A) \to \mathrm{Proj}_2(A) \to \mathrm{Proj}_3(A) \to \ldots$$

by the map: $p \in \mathrm{Proj}_n(A) \to \mathrm{diag}(p, 0) \in \mathrm{Proj}_{n+1}(A)$. We have also the sequence of embeddings;

$$U_1(A) \to U_2(A) \to U_3(A) \to \ldots$$

by the map: $u \in U_n(A) \to \mathrm{diag}(u, 1) \in U_{n+1}(A)$. We write 0_n as zero in $M_n(A)$ and 1_n as the identity of $M_n(A)$. From our discussion before three equivalences among projections eventually coincide in these embeddings. Let $D_n(A)$ be the unitary equivalence classes of $\mathrm{Proj}_n(A)$. We then have a sequence of embeddings $\{D_n(A), \varphi_n\}$ with embedding morphism φ_n (which is not necessarily injective). Put

$$D(A) = \lim_{\rightarrow} D_n(A) \quad ,$$

and write the class $[p]$ in $D(A)$ for a projection p in $M_n(A)$. Usually the sum of two projections is not a projection but we can define the sum in $D(A)$ as the induced operation by the map;

$$(p, q) \in \mathrm{Proj}_n(A) \times \mathrm{Proj}_m(A) \rightarrow \mathrm{diag}(p, q) \in \mathrm{Proj}_{n+m}(A) \quad .$$

Since $\mathrm{diag}(p, q)$ and $\mathrm{diag}(q, p)$ are unitarily equivalent, the set $D(A)$ becomes an abelian semigroup.

<u>Definition 5.2.2.</u> The K_0-group a unital C*-algebra A is the universal group of $D(A)$, that is $K_0(A) = \mathscr{U}(D(A))$.

Roughly speaking, $K_0(A)$ is the universal group of the semigroup of equivalence classes of projections in $M_\infty(A) = \bigcup_{n=1}^{\infty} M_n(A)$. Thus one may also take, as the basic semigroup, the equivalence classes of projections in the C*-tensor product $A \otimes C(H)$. By definition one sees that $K_0(C) = K_0(M_n) \cong Z$. Moreover, we have that $K_0(A) = K_0(M_n(A))$ for every n.

Let ρ be a *-homomorphism of a C*-algebra A to a C*-algebra B. Then it naturally induces a *-homomorphism $\rho_n : |a_{ij}| \in M_n(A) \rightarrow [\rho(a_{ij})] \in M_n(B)$, which sends equivalent projections in $M_n(A)$ to equivalent projections in $M_n(B)$. Therefore, the map ρ defines finally the (semigroup) homomorphism ρ'_* of $D(A)$ to $D(B)$. Hence by the universal property of $\mathscr{U}(D(A))$ we obtain the homomorphism ρ_* of $K_0(A)$ to $K_0(B)$.

Now let A be a non-unital C*-algebra and A_1 be the C*-algebra with identity adjoined. Then the quotient map $q : A_1 \rightarrow A_1/A \cong C$ is a homomorphism of A_1 and defines the induced homomorphism q_* of $K_0(A_1)$ to $K_0(A_1/A) \cong Z$. The K_0-group of A is then defined as the kernel of this homomorphism q_*. We remark that the above functorial property of the K_0-group is still valid for non-unital case.

Next let $U_n(A)_0$ be the connected component of the identity in $U_n(A)$. Then corresponding to the sequence of embeddings of $\{U_n(A)\}$

we have the sequence of embeddings of factor groups;

$$U_1(A)/U_1(A)_0 \to U_2(A)/U_2(A)_0 \to \cdots \quad .$$

<u>Definition 5.2.3.</u> $K_1(A) = \lim\limits_{\to} U_n(A)/U_n(A)_0$.

From the remark after (5.2.3) we see that $K_1(A)$ is an abelian group. However the group $U_1(A)/U_1(A)_0$ need not be abelian in general, hence the above embedding maps are not necessarily injective.

Let $GL_n(A)$ and $GL_n(A)_0$ be the group of invertible elements of $M_n(A)$ defined before and the connected component of the identity in $GL_n(A)$ respectively. Then since $GL_n(A)/GL_n(A)_0 = U_n(A)/U_n(A)_0$, the group $K_1(A)$ is defined also as the limit group of $GL_n(A)/GL_n(A)_0$ without using unitary groups. This corresponds to the fact that $K_0(A)$ may be defined by using idempotents rather than projections, and they would suggest the K-theory for Banach algebras.

Now it is easy to see that $K_1(C) = \{0\}$. Furthermore, the K_1-group of a von Neumann algebra is trivial because spectral resolutions of unitary operators of a von Neumann are obtained within the algebra.

Let ρ be a unital *-homomorphism of a C*-algebra A to a C*-algebra B, then ρ induces a sequence of group homomorphism of $\{U_n(A)\}$ to $\{U_n(B)\}$. Hence we get a sequence of group homomorphisms $\{\rho_{n*}\}$ of $\{U_n(A)/U_n(A)_0\}$ to $\{U_n(B)/U_n(B)_0\}$, finally inducing a homomorphism ρ_* from $K_1(A)$ to $K_1(B)$.

The K_1-group of a non-unital C*-algebra A is defined as in the case of K_0-group. Since, however, $K_1(C) = \{0\}$, $K_1(A)$ coincides with $K_1(A_1)$. Then for a *-homomorphism ρ from a C*-algebra A to a C*-algebra B we also get the induced homomorphism $\rho_*: K_1(A) \to K_1(B)$ by considering the extended unital homomorphism ρ_1 from A_1 to B_1. Here we note that even when A is already unital we can still define the C*-algebra A_1 adjoined the identity giving the product C*-algebra structure in $A_1 = A \oplus \mathbb{C}(1-e)$ where e is a unit of A.

The group $K_0(A)$ coincides with the algebraic K_0-group of A as a ring. The latter is defined as the universal group of the abelian semigroup consisting of the isomorphism classes of finitely generated projective modules over A. These modules are represented by idempotent elements in $M_\infty(A)$ and module isomorphisms among them are reduced to the equivalence relation among idempotent elements. However the present K_1-group does not coincide with the algebraic K_1-group because the notion of the connected component of the identity, $U_n(A)_0$, is not compatible with the algebraic context. Next let X be a compact space and let V be a (complex) vector bundle over X. It is then known that there exists a complementary bundle V^c such that the sum $V \oplus V^c$ becomes a product bundle of some dimension n. This means that V is represented by a projection valued continuous function p(x) from X into $M_n(C)$. Since the C*-algebra of all $M_n(C)$-valued continuous functions on X, $C(X, M_n(C))$ coincides with the algebra $C(X) \otimes M_n = M_n(C(X))$, we may regard the function p(x) as a projection p in $M_n(C(X))$. Moreover, as in the case of algebraic K_0-group, we see that isomorphic vector bundles correspond to equivalent projections (as idempotent elements, hence as projections by (5.2.6)). Since the K_0-group of the space X is defined as the universal group of the semigroup of equivalence classes of vector bundles over X, $K_0(C(X))$ naturally coincides with the K_0-group of the compact space X. Thus, the K_0-group of C*-algebras is quite a natural setting absorbing both algebraic and topological features of C*-algebras.

Proposition 5.2.9. Let A be a (unital) C*-algebra and let $\{A_\alpha\}$ be an increasing net of C*-subalgebras of A such that $A = \overline{\bigcup_\infty A}$, then $K_0(A) = \varinjlim K_0(A_\alpha)$.

For the proof we note that if p is a projection of A there is a projection q in some A_α such that $\|p - q\| < 1$. Therefore we can use (5.2.7). Indeed, take a selfadjoint element a in some C*-subalgebra A_α with $\|p - a\| < \frac{1}{2}$, then putting $\varepsilon = \|p - a\|$ we have,

$$\mathrm{sp}\,a \subseteq [-\varepsilon, \varepsilon] \cup [1-\varepsilon, 1+\varepsilon] \quad .$$

Let $\chi(t)$ be the characteristic function of the interval $[\frac{1}{2}, 1+\varepsilon]$, then since the restriction $\chi(t)|\mathrm{sp}\,a$ is continuous we obtain a projection $q = \chi(a)$ in A_α. By definition, $\|p - q\| < 1$.

Now we shall consider higher K-groups of C*-algebras. The context will proceed as an algebraic transplantation of topological K-theory. Thus we define the suspension SA as the algebra of all A-valued continuous functions on R vanishing at infinity. Denote the algebra $C(S^1, A)$ by ΩA. The algebras SA and ΩA may also be regarded as the algebras of A-valued functions on $[0, 1]$ with $f(0) = f(1) = 0$ and $f(0) = f(1)$ respectively. Then following the split exact sequence $0 \to SA \to \Omega A \overset{\eta}{\to} A \to 0$ where η is evaluation at 1 we obtain a split exact sequence

$$0 \to K_1(SA) \to K_1(\Omega A) \overset{\eta_*}{\to} K_1(A) \to 0 \quad ,$$

so that $K_1(SA) = \eta_*^{-1}(0)$. With this picture in mind we have

<u>Theorem 5.2.10.</u> (1) There is an isomorphism θ_A between $K_1(A)$ and $K_0(SA)$ such that, whenever ρ is a *-isomorphism of A to A, the following diagram commutes;

$$\begin{array}{ccc} K_1(A) & \xrightarrow{\rho_*} & K_1(B) \\ \theta_A \downarrow & & \downarrow \theta_B \\ K_0(SA) & \xrightarrow{S\rho_*} & K_0(SB) \end{array}$$

(2) There is an isomorphism β_A between $K_0(A)$ and $K_1(SA)$ such that the following diagram commutes;

$$\begin{array}{ccc} K_0(A) & \xrightarrow{\rho_*} & K_0(B) \\ \beta_A \downarrow & & \downarrow \beta_B \\ K_1(SA) & \xrightarrow{S\rho_*} & K_1(SB) \end{array}$$

Setting $S^n(A) = S(S^{n-1}A)$, we can define higher K-groups as $K_2(A) = K_0(S^2A)$, $K_3(A) = K_0(S^3A)$ and so on. The above results, however, show that the so-called Bott periodicity $K_n(A) \cong K_{n+}(A)$ also holds for our K-theory.

We have the standard six-term exact sequence.

<u>Theorem 5.2.11.</u> Let $0 \to J \overset{L}{\to} A \overset{q}{\to} A/J \to 0$ an exact sequence for a closed ideal J of a C*-algebra A. Then there are connecting maps ∂_i from $K_0(A/J)$ to $K_1(J)$ and from $K_1(A/J)$ to $K_0(J)$ respectively such that the following standard six-term cyclic sequence is exact;

$$\begin{array}{ccccc} K_0(J) & \xrightarrow{L_*} & K_0(A) & \xrightarrow{q_*} & K_0(A/J) \\ \partial_1 \uparrow & & & & \downarrow \partial_2 \\ K_1(A/J) & \xleftarrow{q_*} & K_1(A) & \xleftarrow{L_*} & K_1(J) \end{array} .$$

In these arguments; the homotopy invariance of K-groups plays a crucial rôle. Namely,

"If ρ_0 and ρ_1 are *-homomorphisms of a C*-algebra A to a C*-algebra B which are homotopic, then ρ_0 and ρ_1 induce the same homomorphism

$$\rho_{0*} = \rho_{1*} : K_*(A) \to K_*(B)" \quad .$$

This will be seen from our previous results.

K-theory of C*-algebras appear in our context through the following results.

<u>Theorem 5.2.12.</u> (Pimsner-Voiculescu). Let A be a C*-algebra with a *-automorphism α. Then there are connecting maps δ_i from $K_1(A \rtimes_\alpha Z)$ to $K_0(A)$ and from $K_0(A \rtimes_\alpha Z)$ to $K_1(A)$ such that the following six-term cyclic sequence is exact;

$$\begin{array}{ccccc}
K_0(A) & \xrightarrow{id_* - \alpha_*} & K_0(A) & \xrightarrow{L_*} & K_0(A \rtimes_\alpha Z) \\
\delta_1 \uparrow & & & & \downarrow \delta_2 \\
K_1(A \rtimes_\alpha Z) & \xleftarrow{L_*} & K_1(A) & \xleftarrow{id_* - \alpha_*} & K_1(A)
\end{array}$$

Theorem 5.2.13. (Connes' Thom isomorphism theorem) If α is an action of R on a C*-algebra A, then

$$K_i(A \rtimes_\alpha R) = K_{1-i}(A) \qquad (i = 0, 1) \quad .$$

Thus, in particular, $K_1(C(X) \rtimes_\alpha R)$ reduces to the K_0-group of the compact space X.

Before showing an example of applications of K-groups in our context, we finally mention the order structure of $K_0(A)$ and the axiomatic K-theory of C*-algebras. Let A be a unital C*-algebra and denote $K_0(A)_+$ the image of $Proj(M_\infty(A))$ in $K_0(A)$. Then we see that

$$K_0(A) = K_0(A)_+ - K_0(A)_+$$

and

$$K_0(A)_+ \cap (-K_0(A)_+) = \{0\} \quad ,$$

which means that we may regard $K_0(A)$ as an ordered group with the positive portion $K_0(A)_+$. This order structure often plays an important rôle in the K-theory of C*-algebras. For instance, when A is an AF-algebra the ordered group $(K_0(A), K_0(A)^+)$ is shown to be complete algebraic invariant for A. In this case, this ordered group is called the dimension group of A.

Now let E be a covariant functor defined on C*-algebras with values in abelian groups. Following basic properties of K_0-group we consider the axioms (A1)-(A5) for the functor E.

(A1) Homotopy invariance as cited before for $K_0(B)$.

(A2) Half exactness: If $0 \to J \overset{i}{\to} A \overset{\rho}{\to} B \to 0$ is an exact sequence of C*-algebras, then $E(J) \xrightarrow{i_*} E(A) \xrightarrow{\rho_*} E(B)$ is exact.

(A3) Stable property: $E(A) = E(A \otimes C(H))$.

(A4) Continuity as cited in (5.2.9) for $K_0(A)$.

(A5) Normalization: $E(C) \cong Z$, $E(C_0(R)) = \{0\}$.

It is then known that these axioms determine, to a considerable extent, the functor K_0 as well as other functors of K-groups. Actually a covariant functor E satisfying (A1)-(A5) coincides with K-functor i.e. $E_*(A) = K_*(A)$ where $E_i(A) = E(S^i A)$ on a comparatively wide class of C*-algebras such as the class of separable C*-algebras of type I together with their C*-inductive limits (cf. Ref. 4). This means that we sometimes need not be concerned with definitions of K-groups in calculation of those groups.

For our later use we shall calculate K-groups of $C(T^n)$ or T^n where T^n means the n-dimensional torus. We assume however the first facts that $K_0(C(T)) = K_0(T) \cong Z$ and $K_1(C(T)) \cong Z$. Now suppose that the group Z acts trivially on a C*-algebra A. Then the crossed product $A \rtimes_{\alpha} {}_r Z$ coincides with the C*-tensor product $A \otimes C(T)$. Since, in this case, $id_* - \alpha_* = 0$ the Pimsner-Voiculescu six-term exact sequence reads;

$$\begin{array}{ccccc}
K_0(A) & \xrightarrow{\text{0-map}} & K_0(A) & \xrightarrow{L_*} & K_0(A \otimes C(T)) \\
\uparrow{\scriptstyle \delta_1} & & & & \downarrow{\scriptstyle \delta_2} \\
K_1(A \otimes C(T)) & \xleftarrow{L_*} & K_1(A) & \xleftarrow{\text{0-map}} & K_1(A)
\end{array}$$

If $A = C(T)$, then $A \otimes C(T) = C(T^2)$. Thus we may consider that it consists of two short exact sequences:

$$0 \to Z \to K_0(C(T^2)) \to Z \to 0 \quad ,$$

$$0 \to Z \to K_1(C(T^2)) \to Z \to 0 \quad .$$

As these sequences split, we obtain

$$K_0(C(T^2)) \cong Z^2 \quad , \quad \text{and} \quad K_1(C(T^2)) \cong Z^2 \quad .$$

Similarly applying the above argument for the algebra $A = C(T^n)$ we get the following

Proposition 5.2.14.

$$K_0(C(T^{n+1})) \cong Z^{2^n} \quad , \quad K_1(C(T^{n+1})) \cong Z^{2^n} \quad .$$

An irrational rotation map σ_θ on T is homotopic to the identity homeomorphism, so that the induced automorphism α_θ is also homotopic to the identity. It follows that $(\alpha_\theta)_* = id_*$ on the K-groups of $C(T)$, and we obtain the same diagram as above for the Pimsner-Voiculescu exact sequence. Consequently, we have $K_0(A_\theta) \cong Z^2$ and $K_1(A_\theta) \cong Z^2$.

5.3 Transformation Group C*-algebras Associated with Subgroups of the Torus T

In this section we shall study transformation group C*-algebras arising from topological dynamical systems associated with subgroups of T. We denote T_d the torus with discrete topology and consider a subgroup of T_d. Let $u : T \to \mathcal{D}$ be the unitary representation on H defined as

$$u(t)\xi_n = e^{2\pi int}\xi_n \quad .$$

Namely, for each $t \in T$, $u(t)$ is a diagonal operator. Let G be a subgroup of T_d, then

$$u(G) = \{u(g) \mid g \in G\} \subset \mathcal{D} \quad .$$

We consider this set as the set $W(\mathcal{S})$ discussed in Sec. 5.1 and write $A_G = C^*(u(G))$. The C*-algebra A_G is commutative and

$$su(g)s^* = e^{-2\pi ig}u(g) \in A_G \quad .$$

Hence, $\{A_G, \text{ad } s\}$ is a covariant system and it determines canonically a shift dynamical system Σ_G. The following theorem characterizes this system. Before going to state the theorem we must however recall that a compact abelian group is monothetic if and only if it is the dual group of a subgroup of T_d (cf. Appendix A).

Theorem 5.3.1. Let G be an infinite subgroup of T_d. Then the system $\Sigma_G = (X, \sigma, \phi)$ is strictly conjugate to the shift dynamical system $(\hat{G}, \sigma_G, \phi_G)$ for a monothetic group $\hat{G}$ where $\sigma_G(\gamma) = \gamma + \phi_G(1)$ and the map $\phi_G : Z \to \hat{G}$ is defined as $< g, \phi_G(n) > = e^{2\pi i n g}$.

Proof. Let B be the linear span of $u(G)$, which is a dense *-subalgebra of A_G. Define the map $\Phi : B \to C(\hat{G})$ by

$$\Phi\left(\sum_{k=1}^{p} \alpha_k u(g_k)\right)(\gamma) = \sum_{k=1}^{p} \alpha_k < g_k, \gamma > \quad .$$

The map Φ is well-defined and isometric. Indeed,

$$\left\| \Phi\left(\sum_{k=1}^{p} \alpha_k u(g_k)\right)\right\| = \sup_{\gamma \in \hat{G}} \left| \sum_{k=1}^{p} \alpha_k < g_k, \gamma > \right|$$

$$= \sup_{n} \left| \sum_{k=1}^{p} \alpha_k < g_k, \phi_G(n) > \right|$$

$$= \sup_{n} \left| \sum_{k=1}^{p} \alpha_k e^{2\pi i n g_k} \right|$$

$$= \left\| \sum_{k=1}^{p} \alpha_k u(g_k) \right\| \text{ (operator norm)} \quad .$$

Moreover, as $u(g)^* = u(-g)$, Φ is an isometric *-isomorphism from B into $C(\hat{G})$ such that the image $\Phi(B)$ separates elements of $\hat{G}$. Therefore $\Phi(B)$ is dense in $C(\hat{G})$ by the Stone-Weierstrass theorem and Φ extends to a *-isomorphism between A_G and $C(\hat{G})$. Let h be the homeomorphism of X to $\hat{G}$ induced by Φ, that is, $\hat{a}(h^{-1}\gamma) = \Phi(a)(\gamma)$ where $\hat{a}$ means the Gelfand representation of $a \in A_G$. Then for every $g \in G$ we have

$$\widehat{u(g)}(h^{-1}\phi_G(n)) = \Phi(u(g))(\phi_G(n)) = e^{2\pi ing}$$

$$= \text{coefficient of } u(g) \text{ at } \xi_n$$

$$= \widehat{u(g)}(\phi(n)) \quad .$$

Hence $h^{-1}\phi_G(n) = \phi(n)$, i.e. $h\phi(n) = \phi_G(n)$. This means that $h \cdot \sigma = \sigma_G \cdot h$ on $\phi(Z)$, whence $h \cdot \sigma = \sigma_G \cdot h$ on X. Thus, Σ_G is strictly conjugate to the dynamical system $(\hat{G}, \sigma_G, \phi_G)$.

Let G be a subgroup of T_d. Since the dual group $\hat{G}$ has a unique σ_G-invariant faithful measure and the system $(\hat{G}, \sigma_G)$ is minimal, the C*-algebra $C^*(\Sigma_G)$ is a simple C*-algebra with unique faithful tracial state. Moreover the dynamical system Σ_G is equicontinuous and minimal, hence by (1.2.1) it is strictly ergodic.

We shall illustrate examples of important compact monothetic groups as shift dynamical systems.

1. $G = \{\frac{k}{p^n} \mid k = 0,1,2,\ldots,p^n - 1,\ n \in Z\}$ for a prime number p.

The dual $Z_p = \hat{G}$ is then the group of p-adic numbers which is homeomorphic to the Cantor set. If p runs over all prime numbers, the dual $\hat{G}$ becomes the product group $\prod_p Z_p$ called the profinite group. Since both groups contain naturally the integer group Z as a dense subgroup, $\phi(n)$ is just the integer in Z. The algebra $C(Z_p)$ is the closure of the union of all periodic sequences of period p^n $(n = 1,2,3,\ldots)$, whereas $C(\pi Z_p)$ is the closure of the union of all periodic sequences for all periods.

2. $G = \{n\theta \mid n \in Z\}$ for an irrational number $\theta \in T$. In this case, $\hat{G} = X$ is obviously the torus T. If we consider (rationally) independent irrational numbers $\theta_1, \theta_2, \ldots, \theta_n$, then the resulting monothetic group (the dual) is just the n-dimensional torus with the translation homeo-

morphism,

$$(t_1, t_2, \ldots, t_n) \to (t_1 + \theta_1, t_2 + \theta_2, \ldots, t_n + \theta_n) \quad .$$

3. $G = \{\frac{k}{p} + k_1\theta_1 + k_2\theta_2 + \ldots + k_q\theta_q \mid k = 0,1,\ldots,p-1;$

$k_1, k_2, \ldots, k_q \in Z\} \quad .$

The group G is of the form $Z/pZ \times T^q$, the p-pieces of q-dimensional torus. We have here the homeomorphism,

$$(n, (t_1, t_2, \ldots, t_q)) \to (n + 1(\text{mod } p), (t_1 + \theta_1, \ldots, t_q + \theta_q)) \quad .$$

4. If we take T_d itself the dual $\hat{G} = Z_b$ is the Bohr compactification of Z, which is a totally disconnected space.

Let τ be the unique tracial state on $C^*(\Sigma_G)$, then τ induces a natural homomorphism $\hat{\tau}$ of the K_0-group of $C^*(\Sigma_G)$ into the real number field R. Denote by $\mathscr{R}_\tau(C^*(\Sigma_G))$ the image of K_0-group by $\hat{\tau}$. The following result shows that this set completely determines the C*-algebra $C^*(\Sigma_G)$. We write

$$G^\downarrow = \{e^{2\pi it} \mid t \in G\}$$

for a subgroup G of T_d.

<u>Theorem 5.3.2.</u> Let G be an infinite subgroup of T_d. Then

(1) $C^*(\Sigma_G)$ is a simple C*-algebra with unique tracial state τ;

(2) $\mathscr{R}_\tau(C^*(\Sigma_G)) = \{t \in R \mid e^{2\pi it} \in G^\downarrow\}$.

The theorem implies that $C^*(\Sigma_{G_1})$ is *-isomorphic to $C^*(\Sigma_{G_2})$ if and only if $G_1 = G_2$.

Although the above second assertion is quite substantial in our context it is beyond the size of this text to give the detailed proof. Thus we give here merely a flavor of the proof.

We first take a monotone increasing net of finitely generated subgroups of G, $\{G_\lambda\}$ such that $G = \bigcup_\lambda G_\lambda$. Then $C^*(\Sigma_{G_\lambda})$ may be regarded as a C*-subalgebra of $C^*(\Sigma_G)$ and the family $\{C^*(\Sigma_{G_\lambda})\}$ forms an increasing net of C*-subalgebras of $C^*(\Sigma_G)$. We then have

$$C^*(\Sigma_G) = \overline{\bigcup_\lambda C^*(\Sigma_{G_\lambda})} \quad .$$

It follows that $K_0(C^*(\Sigma_G)) = \varinjlim K_0(C^*(\Sigma_{G_\lambda}))$ by (5.2.9). Thus putting $\tau_\lambda = \tau | C^*(\Sigma_{G_\lambda})$, we obtain

$$\mathscr{R}_\tau(C^*(\Sigma_G)) = \bigcup_\lambda \mathscr{R}_{\tau_\lambda}(C^*(\Sigma_{G_\lambda})) \quad .$$

Therefore the proof of the theorem is reduced to the case of a finitely generated subgroup G, and this is a hard part of the proof. When G is a finite group, we may write

$$G = \{\frac{k}{n} \mid k = 0,1,2,\dots,n-1\} \quad .$$

In this case, we have seen in Sec. 5.1 that

$$C^*(\Sigma_G) \cong C(T) \otimes M_n \quad .$$

Hence,

$$K_0(C^*(\Sigma_G)) = K_0(C(T)) = K_0(T) \cong Z \quad .$$

On the other hand, since the range of the normalized trace for M_n on projections is the set $\{0, \frac{1}{n}, \frac{2}{n}, \dots, 1\}$ we have

$$\mathscr{R}_\tau(C^*(\Sigma_G)) = \{t \in R | e^{2\pi i t} \in G^\perp\} \quad .$$

The starting point in the case of a finitely generated infinite subgroup is the case of an infinite cyclic group, that is, $G = \{n\theta\}$ for an irrational number θ. In this case $C^*(\Sigma_G)$ is *-isomorphic to the algebra A_θ and the result follows from (4.3.4) and the fact that $K_0(A_\theta) \cong Z^2$ mentioned at the end of Sec. 5.2. Namely, we see that

$\mathscr{R}_\tau(A_\theta) = Z + Z\theta$ and $K_0(A_\theta)$ and the group $Z + Z\theta$ are isomorphic as ordered groups.

For two irrational numbers θ_1 and θ_2 in $(0, \frac{1}{2})$ we now know that if A_{θ_1} is *-isomorphic to A_{θ_2} then $\theta_1 = \theta_2$. On the other hand, σ_θ and $\sigma_{1-\theta}$ are conjugate by the homeomorphism $\rho(x) = -x$ in T. Therefore, A_{θ_1} and A_{θ_2} are *-isomorphic if and only if $\theta_1 = \theta_2$ or $\theta_1 = 1 - \theta_1$.

For a shift dynamical system Σ the $C^*(\Sigma)$ is not necessarily *-isomorphic to the C*-algebra $C^*(\Sigma_G)$ for a subgroup G of T_d. For instance, Furstenberg[10] shows that for some irrational number θ and a continuous function $f(x)$ on T the homeomorphism σ on the 2-torus T^2 defined as $\sigma(x, y) = (\sigma_\theta x, f(x) + y)$ is minimal but not uniquely ergodic. It follows that tracial states on the associated C*-algebra $C^*(\Sigma)$ are not unique, contrary to the case of $C^*(\Sigma_G)$.

Theorem 5.3.4. For a shift dynamical system $\Sigma = (X, \sigma, \phi)$ the following assertions are equivalent;

(1) Σ is minimal and equicontinuous;

(2) Σ is conjugate to the system Σ_G for a subgroup G of T_d;

(3) Σ is strictly conjugate to the system Σ_G for a subgroup G of T_d;

(4) $C^*(\Sigma) = C^*(\Sigma_G)$ for a subgroup G of T_d.

Proof. Assume the assertion (1), then by (1.2.6) the Ellis semigroup $E(\Sigma)$ becomes a compact monothetic group. Hence, $E(\Sigma) = \hat{G}$ for a subgroup G of T_d and if we define the map T_σ as $T_\sigma \cdot \rho = \sigma \cdot \rho$ and $\psi(n) = \sigma^n$ the shift dynamical system $\Sigma' = (E(\Sigma), T_\sigma, \psi)$ is regarded as the shift dynamical system $\Sigma_G = (\hat{G}, \sigma_G, \phi_G)$. On the other hand, in (1.2.6) the homeomorphism Φ between $E(\Sigma)$ and X is given by $\Phi(g) = g(\phi(0))$ for $g \in E(\Sigma)$. Hence,

$$\Phi(\sigma^n) = \sigma^n(\phi(0)) = \phi(n) \quad ,$$

and Σ is strictly conjugate to Σ', the assertion (3). As the implication (3)$\Rightarrow$(1) is trivial the assertions (1) and (3) are equivalent. Therefore, by (5.1.5) and (5.1.6) we have the whole equivalences.

As shown in Example 1.2.1 the Anzai skew product σ of an irrational rotation σ_θ is not equicontinuous, hence the system $\Sigma = (T^2, \sigma)$ is not conjugate to the system Σ_G for a subgroup G. Here, when $\hat{G} = T^2$ the corresponding subgroup G is a subgroup generated by two irrational numbers θ_1 and θ_2. Denote by α the automorphism of $C(T^2)$ induced by σ. We shall show that $C(T^2) \rtimes_{\alpha r} Z$ is even not *-isomorphic to $C^*(\Sigma_G)$ for $G = \{m\theta_1 + n\theta_2 | m, n \in Z\}$. This will be seen from the facts:

$$K_0(C^*(\Sigma_G)) \cong Z^4$$

whereas

$$K_0(C(T^2) \rtimes_{\alpha r} Z) \cong Z^3 \quad .$$

Set $\rho(t, s) = (t + \theta_1, s + \theta_2)$ and observe first that the homeomorphism ρ is homotopic to the identity map by the homotopy path,

$$\rho_r : (t, s) \to (t + r\theta_1, s + r\theta_2) \qquad 0 \le r \le 1 \quad .$$

Let β be the corresponding automorphism of $C(T^2)$ for ρ. It follows that $\beta_* = \mathrm{id}_*$ on the K-groups of $C(T^2)$ and $\mathrm{id}_* - \beta_* = 0$. Thus we have the six-term exact sequence similarly as in Sec. 5.2:

$$\begin{array}{ccccc}
K_0(C(T^2)) & \xrightarrow{\text{0-map}} & K_0(C(T^2)) & \longrightarrow & K_0(C(T^2)) \rtimes_\beta Z) \\
\uparrow \delta_1 & & & & \downarrow \delta_2 \\
K_1(C(T^2) \rtimes_\beta Z) & \longleftarrow & K_1(C(T^2)) & \xleftarrow{\text{0-map}} & K_1(C(T^2))
\end{array}$$

It follows by (5.2.14) that

$$K_0(C(T^2) \rtimes_\beta Z) \cong Z^2 \oplus Z^2 \cong Z^4 \quad .$$

Next consider the homotopy path:

$$\sigma_r : (t, s) \to (t + r\theta, t + s) \quad ,$$

with the corresponding automorphism α_r. We have $\alpha_{r*} = \alpha_{0*}$ on the K-groups of $C(T^2)$. Now the generators of $K_1(C(T^2))$ are those unitary functions, $u(t, s) = e^{2\pi it}$, $v(t, s) = e^{2\pi is}$ and

$$K(C(T^2)) = \{m|u| \oplus n|v| \,|\, m, n \in Z\} \quad .$$

We see that

$$\alpha_0(u)(t, s) = u(t, s - t) = e^{2\pi it} = u(t, s) \quad ,$$

$$\alpha_0(v)(t, s) = e^{2\pi i(s-t)} = \overline{u(t, s)}v(t, s) \quad .$$

Namely, the map α_{0*} sends $[u]$ and $[v]$ to $[u]$ and $-[u] \oplus [v]$ respectively. It follows that α_{0*} may be considered as the matrix $\begin{pmatrix} 1 & -1 \\ 0 & 1 \end{pmatrix}$, whence $id_* - \alpha_{0*} = \begin{pmatrix} 0 & 1 \\ 0 & 0 \end{pmatrix}$. Next consider two pairs of projections (p_1, q_1) and (p_2, q_2) defined as follows;

$$p_1(t, s) = 1 \quad , \quad q_1(t, s) = 0 \quad ,$$

$p_2(t, s) = w(t, s)^* \begin{pmatrix} 1 & 0 \\ 0 & 0 \end{pmatrix} w(t, s)$, $q(t, s) = \begin{pmatrix} 1 & 0 \\ 0 & 0 \end{pmatrix}$, where $w(t, s) = u_t \begin{pmatrix} e^{2\pi is} & 0 \\ 0 & 1 \end{pmatrix} u_t^*$ and u_t is the unitary of $M_2(C(T))$ used in the proof of (5.2.3). It can be shown that $[p_1] \ominus [q_1]$ and $[p_2] \ominus [q_2]$ are generators of $K_0(C(T^2))$. We then have

$$\alpha_0(p_2)(t, s) = w(t,s - t)^* \begin{pmatrix} 1 & 0 \\ 0 & 0 \end{pmatrix} w(t,s - t)$$

$$= w(t, t)^* p_2(t, s) w(t, t) \quad .$$

This however means that $[\alpha_0(p_2)] = [p_2]$ in $K_0(C(T^2))$. As α_{0*}

apparently leaves other classes of projections fixed, it turns out to be the identity map. Thus we have the diagram for α:

$$
\begin{array}{ccccc}
K_0(C(T^2)) & \xrightarrow{\text{0-map}} & K_0(C(T^2)) & \longrightarrow & K_0(C(T^2)\rtimes_{\alpha} Z) \\
\delta_1 \uparrow & & & & \downarrow \delta_2 \\
K_1(C(T^2)\rtimes_{\alpha} Z) & \longleftarrow & K_1(C(T^2)) & \xleftarrow{\begin{pmatrix} 0 & 1 \\ 0 & 0 \end{pmatrix}} & K_1(C(T^2))
\end{array}
$$

Here the kernel of the map $\begin{pmatrix} 0 & 1 \\ 0 & 0 \end{pmatrix}$ consists of those elements $\{m[u] \mid m \in Z\}$. Thus, we obtain a short exact sequence;

$$0 \to Z^2 \to K_0(C(T^2)\rtimes_{\alpha} Z) \to Z \to 0 \quad .$$

This sequence clearly splits and we get the isomorphism,

$$K_0(C(T^2)\rtimes_{\alpha} Z) \cong Z^3 \quad .$$

Similarly we can show that

$$K_1(C(T^2)\rtimes_{\alpha} Z) \cong Z^3 \quad .$$

NOTE TO THE CONTENTS

The contents of Chap. 1 are the minimum preparation of topological dynamics for later discussions. We had to skip many other (potentially) related notions and results. Our definition of topological transitivity is employed in accordance with the usage in case of C*-algebras and moreover in considering to treat non-separable cases. The contents of Chap. 2 are standard.

Transformation group C*-algebras were studied first in a systematic way by Effros-Hahn[6]. Results in Sec. 3.3 such as (3.3.7) are however found in literature under the countability assumptions for the space X and the group G, though G is assumed to be a locally compact group and its topology is used in discussions. Materials in Sec. 4.1 are largely taken from Ref. 32. Though covariance is added, the present definition of the (discrete) induced representation is a direct extension of the classical definition of the induced representation for a subgroup of a finite group. As we have seen, it appears naturally as the GNS-representation of a pure state extension of a character of $C(X)$. The result (4.2.1) is due to Ref. 19. Theorem (4.3.1) may not be found in literature, whereas (4.3.3) is originally proved in Ref. 25. Simplicity of C*-crossed product is an important problem in the theory of C*-algebras. The study of irrational rotation C*-algebras has been initiated by Rieffel[28] with the result (4.3.4) drawing big attentions of many successors. The equivalence of (2) and

(3) is shown in Ref. 5 for a separable but not necessarily commutative C^*-algebra instead of $C(X)$. The contents of 4.4 are refined versions of those results in Ref. 1. Chapter 5 is based on Refs. 18 and 27 except Sec. 5.2. Axiomatic K-theory in Sec. 5.2 is discussed in Ref. 4.

GENERAL REFERENCES

A. Blackadar, B., *K-theory for operator algebras*, (Springer-Verlag, 1986).

B. Brown, J.R., *Ergodic theory and topological dynamics*, (Academic Press, 1976).

C. Dixmier, J., *C*-algebras*, (North-Holland, 1977).

D. Takesaki, M., *Theory of operator algebras I*, (Springer-Verlag, 1979).

E. Pedersen, G.K., *C*-algebras and their automorphism groups*, (Academic Press, 1979).

F. Walter, P., *An introduction to ergodic theory*, (Springer-Verlag, 1982).

REFERENCES

1. Andersen, J. and Paschke, W., *The rotation algebra*, MSRI preprint, (1985).

2. Arveson, W.B. and Josephson, K.B., "Operator algebras and measure preserving automorphisms II", *J. Functional Analysis* 4 (1969), 100-134.

3. Brenken, B.A., "Representations and automorphisms of the irrational rotation algebra", *Pacific J. Math.* 111 (1984), 257-282.

4. Cuntz, J., "K-theory and C*-algebras", *Proc. of Conf. on K-theory, Springer Lect. Note* 1046 55-79.

5. O'Donovan, D.P., "Weighted shifts and covariance algebras", *Trans. Amer. Math. Soc.* 208 (1975), 1-25.

6. Effros, E.G. and Hahn, F., "Locally compact transformation groups and C*-algebras", *Mem. Amer. Math. Soc.* 75 (1967).

7. Effros, E.G. and Shen, C.L., "Approximately finite C*-algebras and continued fractions", *Indiana J. Math.* 29 (1980), 191-204.

8. Ellis, R., "Locally compact transformation groups", *Duke Math. J.* 24 (1959), 119-125.

9. Fathi, A. and Herman, M., "Existence de diffeomorphisms minimaux", *Asterique* 49 (1977), 37-59.

10. Furstenberg, H., "Strict ergodicity and transformations of the torus", *Amer. J. Math.* 83 (1961), 573-601.

11. Glimm, J., "Families of induced representations", *Pacific J. Math.* 12 (1962), 885-911.

12. Gootman, E.C. and Rosenberg, J., "The structure of crossed products of C*-algebras: A proof of the generalized Effros-Hahn conjecture", *Invent. Math.* 52 (1979), 283-298.

13. Green, P., "C*-algebras of transformation groups with smooth orbit space", *Pacific J. Math.* 72 (1977), 71-97.

14. Green, P., "The local structure of twisted covariant algebras", Acta Math. 140 (1978), 191-250.

15. Ji. R., *On the crossed product of C*-algebras associated with Furstenberg transformations on tori*, (Thesis SUNY Stony Brook, 1986).

16. Kawamura, S., "Invariant subspaces of shift operators of arbitrary multiplicty", *J. Math. Soc. Japan* 34 (1982), 339-354.

17. Kawamura, S., Takemoto, H. and Tomiyama, J., "State extentions in transformation group C*-algebras", to appear in Acta Sci. Math., Szeged.

18. Kawamura, S. and Takemoto, H., "C*-algebras associated with shift dynamical systems", *J. Math. Soc. Japan* 36 (1984), 279-293.

19. Kawamura, S., Tomiyama, J. and Watatani, Y., "Finite dimensional irreducible representations of C*-algebras associated with topological dynamical systems", *Math. Scand.* 56 (1985), 241-248.

20. Mackey, G.W., "Induced representations of locally compact groups I", *Ann. Math.* 55 (1952), 101-139.

21. Moore, C.C., "Groups with finite dimensional irreducible representations", *Trans. Amer. Math. Soc.* 166 (1972), 401-410.

22. Packer, J., *K-theoretic invariants for the C*-algebras associated to transformations and induced flows*, preprint.

23. Pimsner, M.V. and Voiculescu, D., "Imbedding the irrational rotation C*-algebra into an AF-algebra", *J. Operator Theory* 4 (1980), 201-210.

24. Pimsner, M.V., "Embedding some transformation group C*-algebras into AF-algebras", *Ergodic Theory and Dynamical Sys.* 3 (1983), 613-626.

25. Power, S.C., "Simplicity of C*-algebras of minimal dynamical systems", *J. London Math. Soc.* 18 (1978), 534-538.

26. Putnam, I., Schmidt, K. and Skau, C., *C*-algebras associated with Denjoy homeomorphisms of the circle*, preprint.

27. Riedel, N., "Classification of the C*-algebras associated with minimal rotations", *Pacific J. Math.* 101 (1982), 153-162.

28. Rieffel, M.A., "C*-algebras associated with irrational rotations", *Pacific J. Math.* 93 (1981), 415-429.

29. Taylor, J.L., *Banach algebras and topology in "Algebras in Analysis"*, (Academic Press, 1975).

30. Takesaki, M., "Covariant representations of C*-algebras and their locally compact automorphism groups", *Acta Math.* 119 (1967), 273-303.

31. Williams, D.P., "The topology on the primitive ideal space of transformation group C*-algebras and C.C.R. transformation group C*-algebras", *Trans. Amer. Math. Soc.* 266 (1981), 335-359.

32. Kawamura, S. and Tomiyama, J., *A class of irreducible representations of transformation group C*-algebras*, preprint.